Atomic Spectra

T. P. Softley
Physical Chemistry Laboratory, University of Oxford

OXFORD NEW YORK TOKYO
OXFORD UNIVERSITY PRESS

Oxford University Press, Walton Street, Oxford OX2 6DP

Oxford New York
Athens Auckland Bangkok Bombay
Calcutta Cape Town Dar es Salaam Delhi
Florence Hong Kong Istanbul Karachi
Kuala Lumpur Madras Madrid Melbourne
Mexico City Nairobi Paris Singapore
Taipei Tokyo Toronto
and associated companies in
Berlin Ibadan

Oxford is a trade mark of Oxford University Press

Published in the United States
by Oxford University Press Inc., New York

First published 1994
Reprinted 1996 (with corrections)

A catalogue record for this book is available from the British Library

Library of Congress Cataloging in Publication Data
Softley, T. P. (Tim P.)
Atomic spectra / T.P. Softley.
(Oxford chemistry primers : 19)
Includes bibliographical references and index.
1. Atomic spectroscopy. I. Title. II. Series.
QD96.A8S58 1994
541.294—dc20 94-20394 CIP

ISBN 0 19 855689 6 (Hbk)
ISBN 0 19 855688 8 (Pbk)

Printed and bound in Great Britain by
The Bath Press, Bath, Somerset

Series Editor's Foreword

Oxford Chemistry Primers are designed to provide clear and concise introductions to a wide range of topics that may be encountered by chemistry students as they progress from the freshman stage through to graduation. The Physical Chemistry series will contain books easily recognized as relating to established fundamental core material that all chemists need to know, as well as books reflecting new directions and research trends in the subject, thereby anticipating (and perhaps encouraging) the evolution of modern undergraduates.

In this fourth Physical Chemistry Primer, Tim Softley has produced an authoritative and beautifully written account of *Atomic Spectra* which covers both elementary and advanced aspects of the subject, and which will be of lasting and broad value throughout any undergraduate's career. This primer will be of interest to all students of chemistry (and their mentors).

Richard G. Compton
Physical Chemistry Laboratory, University of Oxford

Preface

The spectroscopy of atoms has played an indispensable role in the development of quantum mechanical theories of atomic structure, because of the direct information which can be obtained from spectra about atomic energy levels. An understanding of chemical bonding, reactivity, and periodicity can only be built on a secure understanding of electronic structure, and therefore the topic of atomic spectra is very much at the core of any modern chemistry course at university level. This Primer provides a systematic and rigorous introduction to the spectra and electronic structure of atoms, beginning with the hydrogen atom, and following a logical progression though the alkali metals, and the helium atom, to atoms with many unpaired electrons. An emphasis is placed on the explanation of the observed spectra using elementary quantum theory. Experimental methods and modern applications are also described to highlight the importance of atomic spectroscopy in current scientific research and in technological development. Chapters 1 to 3 are written at an introductory level suitable for first-year undergraduates (although some sections, especially 2.5, 2.8, and 2.9 could be deferred until a later course) while the material in chapters 4 and 5 could form the basis for a second-level course.

I am grateful to John Freeman for preparing the many figures of this book and to the third-year undergraduates at Merton (1993–94) who acted as guinea-pigs in reading the book and trying out the problems.

Oxford T.P.S.
May 1994

Contents

1 Quantum mechanics and light

1.1 The purpose and practice of spectroscopy

Spectroscopy involves the investigation of the interaction of light with matter; a *spectrum* shows how this interaction varies with the frequency of the light. Most commonly, either the intensity of light transmitted through the sample is measured as a function of frequency—an *absorption* spectrum—or alternatively the light emitted by a sample, excited in a flame or electric discharge, is analysed for its constituent frequencies—an *emission* spectrum. Figure 1.1 shows schematically the classical method for obtaining an absorption spectrum. The recording of an emission spectrum is obtained using a similar apparatus, but the sample cell is replaced by the flame or discharge, and the white-light source is not needed.

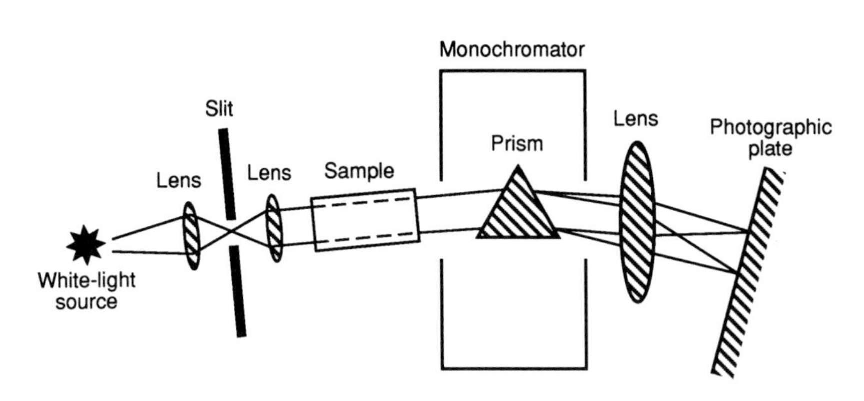

Fig. 1.1 White light, covering a wide range of wavelengths, is passed through the sample and thence to a monochromator. The prism (or alternatively a diffraction grating) *disperses* the light, sending different wavelengths in different directions. A photographic plate then records the intensity of transmitted light as a function of wavelength.

The great value of spectroscopy in chemistry and physics arises from three facts:

1. A spectrum generally shows sharp discrete features, rather than smooth variations with wavelength, as illustrated by the atomic emission spectrum of hydrogen shown in Fig. 1.2a. The discrete stucture is particularly evident when the medium investigated consists of atoms in the gas phase, but also occurs for complex molecules in the gas phase, and sometimes for atoms and molecules in condensed phases too.

2. The discrete features generally appear in well-defined patterns; in most cases the frequencies can be fitted to quite simple mathematical formulae;

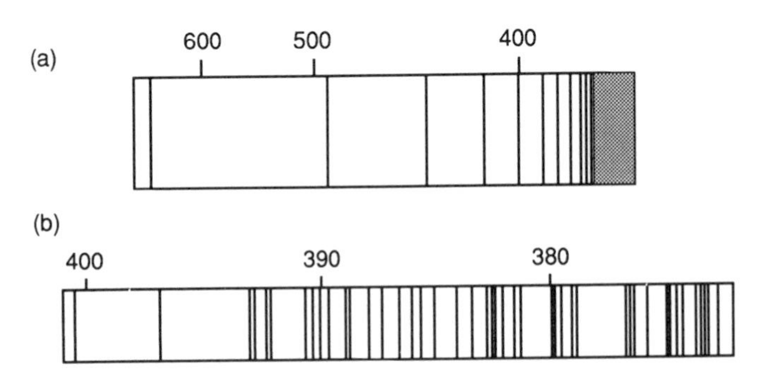

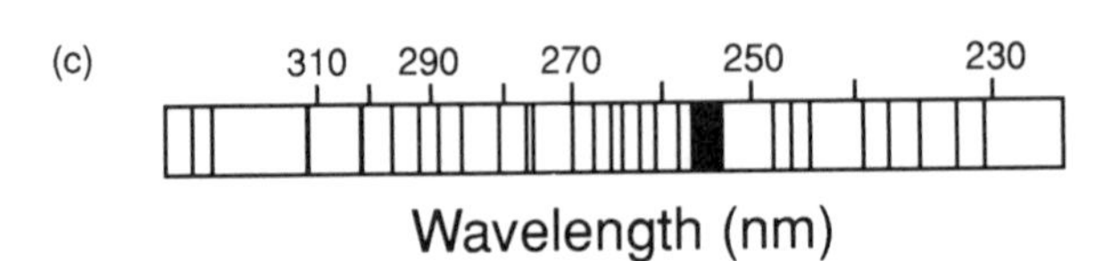

Fig. 1.2 The emission spectra of (a) hydrogen (b) iron and (c) mercury.

for example, all the line frequencies of the Balmer series of hydrogen (Fig. 1.2a) are described by the formula;

$$\nu = a + \frac{b}{n^2}, \tag{1.1}$$

where ν is the frequency of absorption, a, b are constants and n is an integer 3,4,5,. . .

3. The discrete line pattern observed is strongly characteristic of the chemical species producing it, and, even for the most simple atoms, a spectrum is a unique fingerprint that signifies the presence of that species (compare the spectra of hydrogen, iron and mercury in Figs 1.2a, b and c).

The atomic spectrum of hydrogen and the spectra of many other species provided major evidence for the *quantization* of energy. The quantum theory was fully developed in the 1920s and it showed that electrons in atoms and molecules could not exist with any arbitrary energy, but only in a set of *quantized* discrete energy states (or 'levels') that were characteristic of the species in question. The discrete line spectra observed for atoms and molecules reflect the discrete nature of the energy levels. An atom can lose or gain energy by emitting or absorbing light; as the energy states are quantized, the atom can only change energy by making specific energy jumps as illustrated in Fig. 1.3.

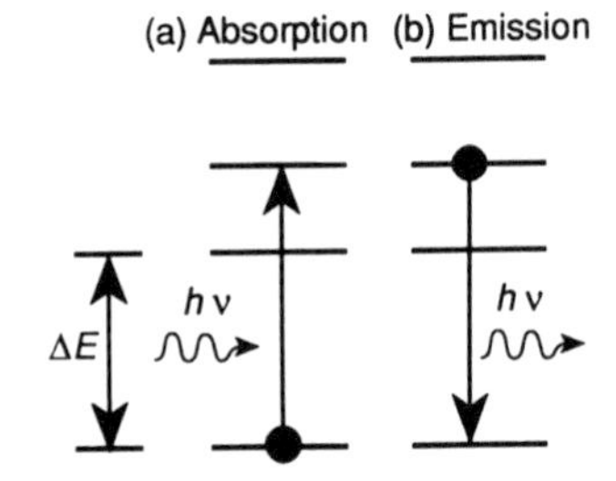

Fig. 1.3 The horizontal lines indicate schematically the quantized energy levels of an atom; in (a) the electron jumps to a higher energy state and absorbs radiation, whereas in (b) it jumps to a lower energy state by emitting radiation.

At the beginning of the twentieth century, Einstein and Planck proposed that there was a direct proportionality between the energy jump ΔE and the frequency of the radiation emitted or absorbed, ν.

$$\Delta E = h\nu. \tag{1.2}$$

h is known as Planck's constant with the value $6.626\,076 \times 10^{-34}$ J s. This simple equation is undoubtedly the most important in spectroscopy, because it provides the connection between the spectrum and the underlying energy levels. By measuring the frequencies of absorption or emission, we can directly determine energy level differences for the atoms being studied.

From the above discussion it should be apparent that two of the primary uses of spectroscopy are as follows.

1. The determination of atomic (or molecular) energy levels. A knowledge of the atomic energy levels is a prerequisite for understanding many aspects of chemical reactivity and structure. Furthermore, the energy levels can be used to test quantum theories and gain a fundamental understanding of the underlying physics.

2. The identification of unknown species by the use of the characteristic fingerprint intrinsic to the spectrum.

It will be seen in subsequent chapters that modern spectroscopy goes well beyond achieving these goals. One application is the determination of concentrations of absorbing/emitting species through measurement of intensities; the distribution of atoms amongst their permitted energy states can usually be obtained in the same way. Atomic spectroscopy underlies the mechanisms of some lasers (Section 1.3), the accurate measurement of frequencies (Section 5.8), the cooling of atoms to submicro-Kelvin temperatures (Section 3.9), and also radio and optical astronomy. Furthermore, the widths of spectral lines can provide dynamical information on lifetimes of excited atoms (Section 5.11).

1.2 The nature of electromagnetic radiation

The classical theory of light

The work of Maxwell, Young, and others in the nineteenth century demonstrated that light could be considered as an *electromagnetic* disturbance propagating through space as a travelling wave; oscillating electric and magnetic fields are directed perpendicular to the direction of propagation. Figure 1.4 shows a snapshot representation of an electromagnetic wave which is travelling forward in the x-direction. The sine curve in the xy-plane represents the magnitude and direction of the electric field, at a given time, t, as a function of position, x. The perpendicular magnetic field, pointing in the z-direction is represented by the shaded sine curve. Mathematically, the magnitude of the electric field E (defined as the electrostatic force on a unit positive charge) is given in the plane wave approximation by the equation,

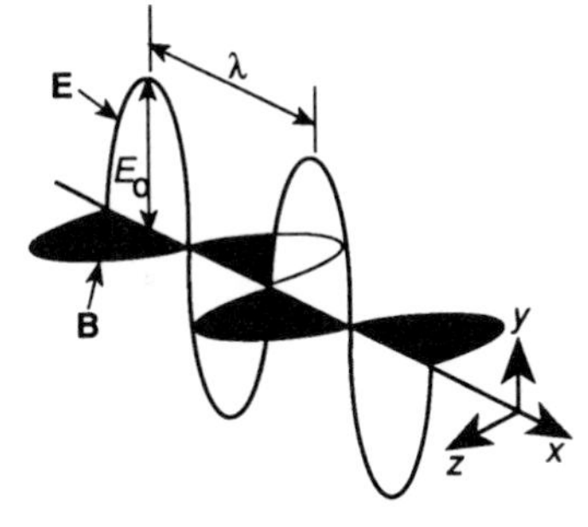

Fig. 1.4 Snapshot representation of a plane electromagnetic wave propagating in the x-direction.

$$E = E_0 \sin\left(\frac{2\pi}{\lambda}(x - vt)\right), \tag{1.3}$$

where E_0 is the amplitude of the wave as indicated in Fig. 1.4. The whole profile shown in the figure moves forward at constant velocity v such that the electric field at given position x varies sinusoidally with time: in a vacuum the velocity is equal to a fundamental constant, $c = 2.997\,924\,58 \times 10^8\,\mathrm{m\,s^{-1}}$. The *wavelength* λ is defined as the distance between successive maxima (or some other pair of equivalent points) on the sine curve, while the *frequency* ν is the number of maxima passing a fixed point in space per second (units Hz $\equiv$ s^{-1}). For light in a vacuum these quantities are related by the fundamental equation

$$\nu = \frac{c}{\lambda}. \tag{1.4}$$

One further quantity often used in spectroscopy is the *wavenumber* $\tilde{\nu}$

$$\tilde{\nu} = \frac{1}{\lambda} = \frac{\nu}{c}, \tag{1.5}$$

for which the non-SI units cm^{-1} are commonly employed. The eqn (1.2) can then be written in the alternative forms,

$$\Delta E = h\nu = \frac{hc}{\lambda} = hc\tilde{\nu}. \tag{1.6}$$

When the wave travels through a medium other than a vacuum, the frequency of the wave is unchanged, but the velocity and wavelength are reduced; the *refractive index* n is defined as the ratio of the velocity of light in a vacuum to that in the medium of interest, v,

$$n = \frac{c}{v} = \frac{\lambda_0}{\lambda} \geq 1, \tag{1.7}$$

where λ_0 is the vacuum wavelength.

For the case where the observer of the light (or the matter interacting with it) is not stationary but moving towards or away from the source of light with relative velocity v_r, the effective frequency ν' detected by the observer is increased or decreased according to

$$\nu' = \nu\left(1 \pm \frac{v_r}{c}\right).$$

This frequency shift is known as the *Doppler effect*. An important consequence is that because atoms in the gas phase will be moving in all directions with a distribution of relative velocities, the discrete absorption features in a spectrum will not be infinitesimally sharp but have a finite linewidth reflecting a range of Doppler shifts, i.e., the atoms do not all absorb at the same source frequency, but at the same Doppler-shifted frequency. In astronomy, the Doppler shift of spectral features in the light emitted by stars can be used to determine the velocity of those stars relative to the Earth.

As an electromagnetic wave travels forward it carries energy with it. The *intensity* is defined as the amount of energy passing through unit area in unit time; the intensity actually oscillates rapidly as the electric field amplitude oscillates, but on average the intensity I is proportional to the square of the electric field amplitude E_0

$$I = \frac{1}{2}\epsilon_0 E_0^2 c, \tag{1.8}$$

($\epsilon_0 = 8.854\,187 \times 10^{-12}\ J^{-1}C^2m^{-1}$ is the permittivity of free space).

The electric and magnetic fields are vector quantities; the *polarization* of the wave refers to the direction of the electric field vector, which is always perpendicular to the direction of travel. Light from an ordinary lamp is *unpolarized* in the sense that the direction of the electric field is rapidly and randomly changing in space (in the yz-plane for a wave propagating in the x-direction), and the direction averages to an isotropic distribution. Light from a laser is normally close to 100% linear polarized implying that the electric field vector always points in one direction.

These simple concepts describe the properties of electromagnetic radiation over an astonshingly wide range of values for the frequency and wavelength. Figure 1.5 illustrates the electromagnetic spectrum, ranging from gamma rays to radio waves, a span of more than 20 orders of magnitude in frequency. For the most part we shall be considering only visible or ultraviolet radiation which is tiny region on the overall picture. Red light in the visible region has a wavelength of about 6×10^{-7} m (600 nm), corresponding to a frequency of 5×10^{14} Hz (500 THz) or a wavenumber of 16 666 cm^{-1}.

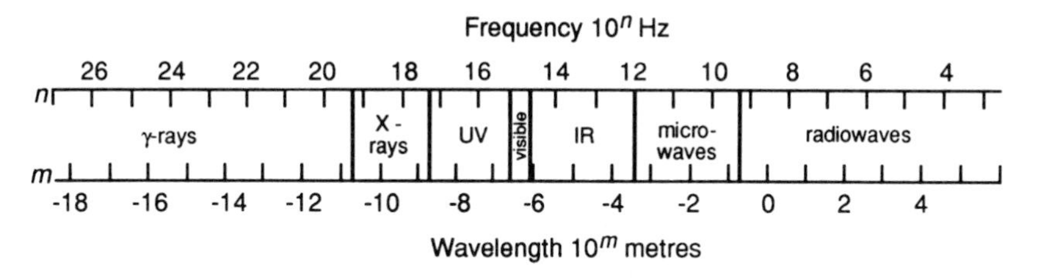

Fig. 1.5 The regions of the electromagnetic spectrum, plotted on a logarithmic scale, as a function of frequency and wavelength.

Photons

Einstein's explanation of the relationship (1.2) was that light was not to be regarded as a continuous electromagnetic field, but was made up of discrete packets of energy known as *photons*. Each photon has an energy

$$E = h\nu. \tag{1.9}$$

A photon is actually a very tiny packet of energy— a photon in the visible spectral region has energy equal to $\sim 3.3 \times 10^{-19}$ J. Thus the number of photons hitting one square meter of the Earth's surface (intensity of sunlight $\sim$ 1 kW m^{-2}) is around 3×10^{21} per second.

Absorption and emission of radiation involves the interaction between individual atoms (or molecules) and individual photons. In the absorption event, the photon gives up its energy to the atom and is effectively destroyed in the process (Fig. 1.3a). The atom becomes excited into a state in which an electron is in a higher energy level. The overall process is subject to conservation of energy, and therefore it is required that the change of energy of the atom ΔE is equal to the energy of the photon, $h\nu$. As there are a restricted number of possible values of ΔE for a given atom with its discrete energy levels, (indeed not all transitions are allowed; see Section 2.7) the absorption occurs at a restricted number of discrete frequencies. Emission is the reverse of the above process, in which an excited atom loses energy ΔE by creating a photon of frequency $h\nu$ (Fig. 1.3b).

The proposal of Planck and Einstein that light consisted of small particle-like photons, is at first sight contradictory to the classical wave description due to Maxwell. However, the fundamental relationship $E = h\nu$ still retains an identification with the wave description of light, as it contains the frequency ν. Furthermore, although the photon description is necessary to describe the absorption and emission of radiation, the wave description appears to be valid and more useful for describing effects such as diffraction and interference. In 1923, de Broglie used Einstein's theory of special relativity, in conjunction with relation (1.9) to show that each photon had a momentum p given by

$$p = \frac{h}{\lambda}. \tag{1.10}$$

Again the wave picture appears through the wavelength λ. The explanation of the apparent wave/particle paradox requires a reinterpretation of what we understand by the electromagnetic wavefunction, eqn (1.3). In the quantum picture of light, we are to consider that the motion of the photons is governed by the wavefunction, such that the probability of finding a photon at a particular point in space is proportional to the square of the electric field amplitude, $E(x, t)^2$. This is not unreasonable, because the classical intensity of light depends on E^2, while in the quantum picture, the rate of transfer of energy is determined by the photon density, which must be proportional to the probability. We will return to this quantum mechanical interpretation of the wavefunction in Section 1.4. But before doing so, we will briefly discuss the properties of a very important type of light source.

1.3 Lasers and stimulated emission

The acronym LASER stands for light amplification by stimulated emission of radiation. A laser is a source of light with very special properties that have made it a device of great technological importance, and one of everyday use (e.g., in compact disc players). These properties may be summarized as follows:

1. **Monochromatic**; most lasers exhibit a high degree of monochromaticity, which means that the range of frequencies emitted by the laser is very narrow. Typically the ratio of the laser frequency bandwidth $\delta\nu$ to the actual laser frequency ν might lie in the range $\delta\nu/\nu = 10^{-5}$ to 10^{-8}. This property is in contrast to classical light sources which generally produce a broad band of frequencies.

2. **Directional**; lasers produce a very well-directed beam of light which under optimum conditions can have a divergence cone angle (see Fig. 1.6) of order λ/d where λ is the wavelength and d the beam diameter. For example, a beam of 2 mm diameter, wavelength 500 nm, travelling a distance of one km would have increased to a diameter of only 50 cm. Lasers can also be focused to a much smaller point than any other type of light source. The minimum size of the focal point is of the order of one wavelength.

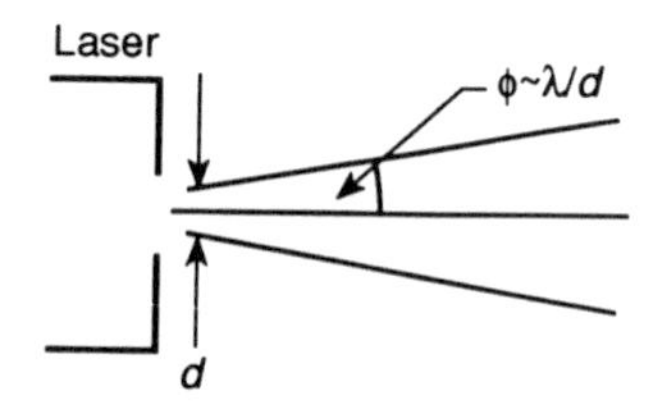

Fig. 1.6 The divergence cone angle, ϕ for a laser beam with initial diameter d.

3. **Coherent**; the radiation emitted from a light source can be considered to be made up of a very large number of wavelets, each of which originates from one particular atom in the source; in effect, each photon is described by its own mathematical wavefunction. In the light emitted by a laser all these wavelets are *in phase* (Fig. 1.7a), meaning that the maxima of all the wavelets occur at the same position in the propagation direction at a given time. This property is called *coherence*. Normal light sources are incoherent meaning that the phases of all the wavelets are random and show no correlation with one another (Fig. 1.7b).

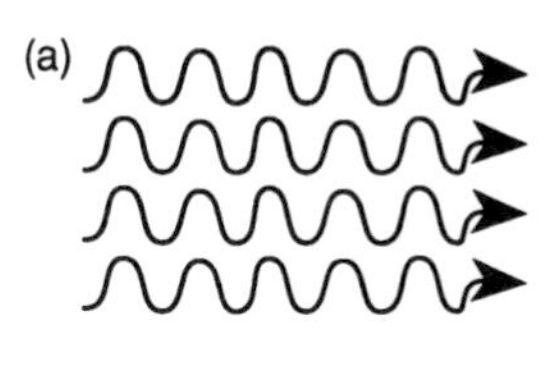

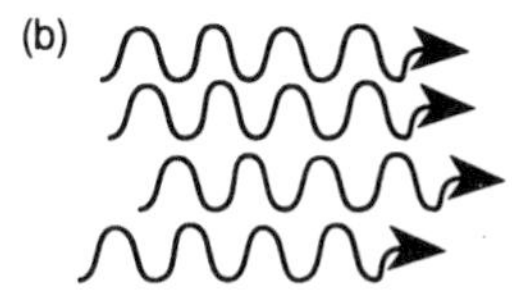

Fig. 1.7 (a) The light from a laser shows a constant phase relationship across a wide area of the beam. (b) The light from an incoherent source shows only random phase relationships, fluctuating with time.

4. **Intense**; some types of lasers are capable of producing light of very high intensity. This is partly a consequence of the amplification that occurs within a laser (see below) but is also associated with the directional and focusing properties (intensity being defined as the amount of light energy passing through unit area in unit time). Furthermore some types of lasers compress all the energy temporally into very short pulses, boosting the peak intensity. For example, a commercially available Nd:YAG laser operating at a wavelength of 1064 nm, can produce peak powers of 10^8 W in a short pulse lasting 10 ns; if focused to a diameter of 1 μm^2, the intensity at the focal point would be 10^{20} W m^{-2}, 10^{17} times more intense than sunlight at the Earth's surface.

5. **Polarized**; the light emitted by a laser is normally linearly polarized as discussed in Section 1.2.

6. **Short temporal pulses**; although certain types of laser are described as *continuous wave*, meaning that they emit light of constant intensity, other types produce short pulses of light at regular intervals. The world record shortest pulse length produced by a laser is currently 4 fs (4×10^{-15} s) although more common laboratory lasers might produce nanosecond (10^{-9} s) or picosecond (10^{-12} s) pulses. A 4 fs pulse in the visible region of the spectrum corresponds to just two cycles of the sinusoidally oscillating electromagnetic field.

The remarkable properties of laser radiation arise largely from the process of *stimulated emission* which lies at the heart of the mode of operation of a laser. We have already discussed briefly the processes of absorption and emission, illustrated in Fig. 1.3. The emission process shown there is more correctly called *spontaneous emission* (or sometimes called fluorescence) in which an atom or molecule that finds itself in a state of high energy spontaneously emits a photon at a random moment in time, and the molecule drops to a lower energy state. Absorption, in which the molecule or atom jumps to a higher energy state, can only occur when a photon of correct frequency interacts with the atom, driving it into the higher state. Stimulated emission (Fig. 1.8) occurs when a photon interacts with an atom in the higher energy state forcing it down to the lower energy state with emission of a *second* photon; the first photon stimulates the emission of the second photon. This process requires both photons to have frequencies obeying the resonance condition $\Delta E = h\nu$. The second photon is *codirectional* with the first photon and is in phase with it, meaning that coherent light is produced.

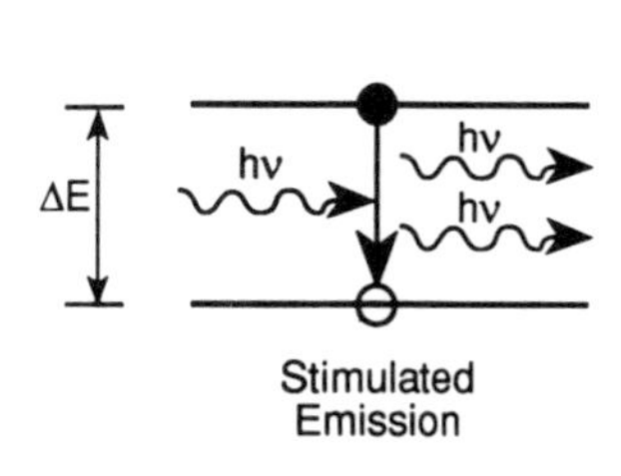

Fig. 1.8 Stimulated emission: the atom is initially in an excited energy state (horizontal line with black dot), but is forced to de-excite by the interaction with a photon obeying the resonance condition $\Delta E = h\nu$.

Inside a laser, a chain of stimulated emission events is set off, each newly generated photon stimulating the emission of a photon from another atom in sequence. There is therefore a build-up of light in the medium, having the properties described above. The maximum amplification can be achieved by putting the medium undergoing stimulated emission between two mirrors as shown in Fig. 1.9, so that the light undergoes many passes back and forth through the medium. One of the mirrors is only partly reflecting, so that some of the coherent light emerges as a highly directional beam as shown.

The mechanism of laser action sounds almost trivial on this description, but we have not considered the fact that many of the generated photons will subsequently interact with an unexcited atom, leading to absorption. Therefore, in order to make a laser work we need to create a medium which has more atoms in an excited energy level than in a lower energy level; this situation is abnormal and is known as a *population inversion*. The mechanisms for producing population inversions are discussed later in Section 4.8; in general an input of energy is required, represented in Fig. 1.9 by the 'pumping process'.

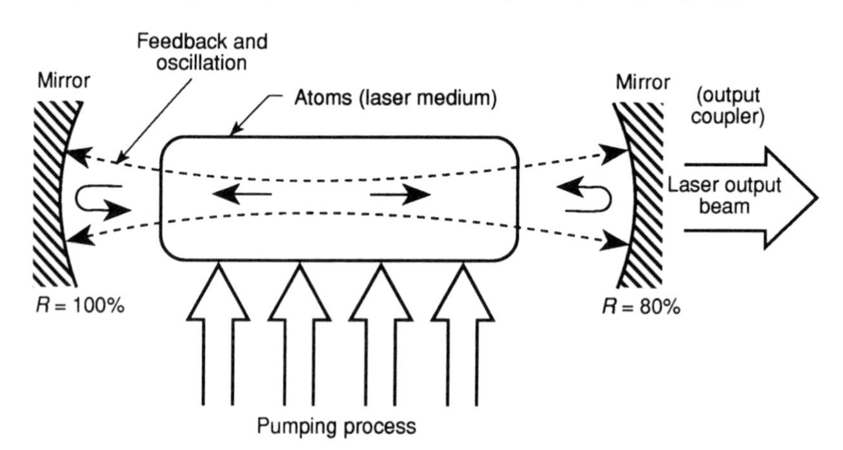

Fig. 1.9 Schematic diagram of a laser.

1.4 Wave mechanical description of an atom

The wavefunction and the Schrödinger equation

In Section 1.2 we discussed the quantum mechanical picture of light, which was orginally thought to behave as a wave, but also could be simultaneously ascribed particle-like properties. The complementary idea proposed by de Broglie in 1923 was that the motion of all particles, e.g., electrons, nucleii, and even macroscopic objects, previously obeying the laws of Newtonian mechanics, should also be represented by wave motion. De Broglie suggested that his relationship (1.10) could be extended to motion of all particles, and therefore he predicted that a beam of electrons should show wave-like properties, such as diffraction at a narrow aperture. This was experimentally demonstrated in 1925 in independent experiments by Davison/Gerner and Thompson/Reid.

Schrödinger concluded that if electrons could be ascribed wave-like properties then their motion must be described using a wavefunction. He set out to show that the waves for electrons in atoms were in some way analogous to the *standing waves* of a vibrating string fixed at both ends (see Fig. 1.10); he presumed that quantization of energy would arise in the same fashion as the quantization of wavelength of the vibrating string (the restriction that the length of string must equal a multiple of $\lambda/2$). It was well known that light waves such as eqn (1.3) obeyed certain second-order differential equations (Maxwell's equations), and, making use of the de Broglie relationship, Schrödinger deduced that particle waves should be solutions of an analogous second-order differential equation of the form

$$\left[\frac{-\hbar^2}{2m}\nabla^2 + V(x, y, z)\right]\Psi = E\Psi \qquad (\hbar = \frac{h}{2\pi}). \tag{1.11}$$

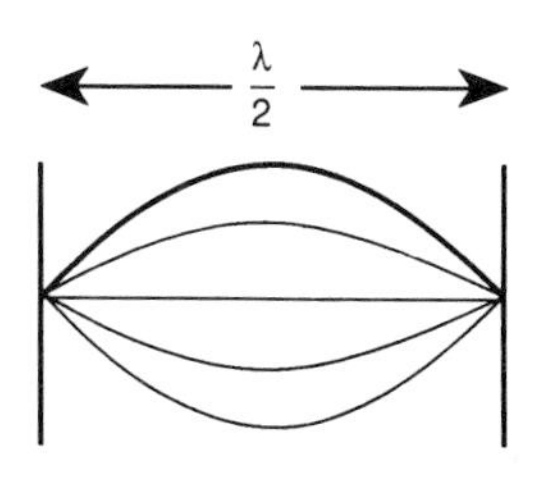

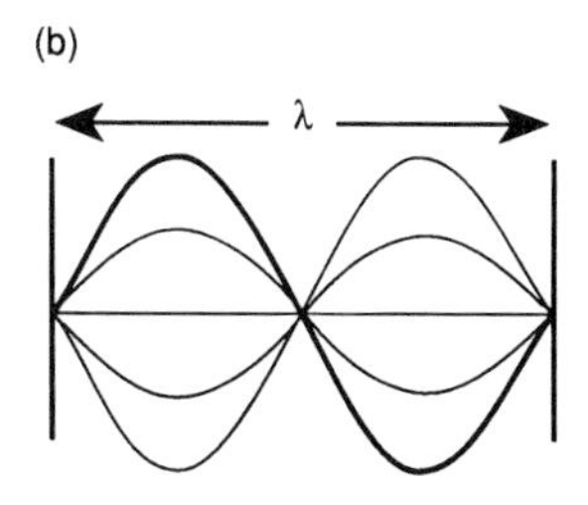

Fig. 1.10 The standing waves of a vibrating string: (a) fundamental mode ($\frac{1}{2}\lambda = l$), (b) first harmonic mode ($\lambda = l$).

Ψ is the wavefunction and is a mathematical function depending on position coordinates x, y, z. E is the total energy of the particle (the sum of its kinetic and potential energies) and is a constant independent of the position coordinates. $V(x, y, z)$ is a function describing how the potential energy of the particle varies with coordinates, and ∇^2 is the differential operator $\frac{\partial^2}{\partial x^2} + \frac{\partial^2}{\partial y^2} + \frac{\partial^2}{\partial z^2}$. The equation is written in a form appropriate for a single particle of mass m; for more than one particle additional terms must be added to the left-hand side as discussed below.

This very important equation, known as the Schrödinger equation, is generally used to obtain the wavefunctions and corresponding energies for a system of interest. It should be noted that every valid wavefunction Ψ corresponds to a specific single value of the energy E. It is seen below that for many situations, there are a restricted number of solutions Ψ of this equation, hence the energy is quantized, i.e., it can only take certain discrete values. This quantization of energy arises primarily as a consequence of *boundary conditions* imposed on the wavefunction as shown below. These conditions are analogous to the requirement for a vibrating string that the amplitude of vibration must be zero at both ends.

The *wavefunction* is a mathematical function from which it is possible to extract information about the position and momentum of a particle or, more generally, of the set of particles comprising a particular body. For example the wavefunction of a $1s$ electron in the hydrogen atom has the form

$$\Psi = \left(\frac{1}{\pi a_0^3}\right)^{\frac{1}{2}} \exp\left(\frac{-r}{a_0}\right), \tag{1.12}$$

where r is the distance of the electron from the nucleus ($r = (x^2 + y^2 + z^2)^{\frac{1}{2}}$) and a_0 is a constant.

In Newtonian mechanics a particle must always have a well-defined position and momentum at a given time: in contrast, one of the major themes of the wave mechanical description of matter is that we can no longer define these quantities with arbitrary position. Instead we can only describe the *probability* of the electron being at a certain position in space, or the probability of it having a certain momentum. The Heisenberg uncertainty principle expresses our inability to specify these quantities precisely through the relationship

$$\Delta p_x \Delta x \geq \frac{\hbar}{2}. \tag{1.13}$$

Here, Δp_x and Δx are the average errors (strictly speaking the rms deviations from the mean) that we would obtain in trying to measure the x-component of the momentum p_x and position x of a particle simultaneously, and therefore represent the uncertainty in these quantities.

The direct link between the wavefunction and probability comes from the *Born interpretation* which in the present context may be expressed as: *The probability of finding an electron in the volume element* $\mathrm{d}\tau$ *at the position (x,y,z) is* $P(x, y, z)\mathrm{d}\tau$ *where*

$$P(x, y, z) = \Psi^*(x, y, z)\Psi(x, y, z). \tag{1.14}$$

$\Psi(x, y, z)$ is the amplitude of the wavefunction describing the motion of the electron at the position x, y, z and Ψ^* is the complex conjugate.

The interpretation immediately imposes certain constraints on the nature of the mathematical function Ψ. First, Ψ must be single-valued and finite otherwise it would not be possible to define a unique value for the probability at a particular position. Secondly, the probability of finding the particle somewhere in space must be unity; therefore, to obey eqn (1.14),

$$\int \Psi^*\Psi \mathrm{d}\tau = 1. \tag{1.15}$$

Thirdly, the requirement for the wavefunction to obey the Schrödinger equation, means that the second derivative of the wavefunction must be well-defined, and therefore the wavefunction cannot undergo sudden jumps or changes of gradient; i.e., it must be *continuous*.

In general, far more information can be extracted from a wavefunction than just the probability distribution for position. This is most easily achieved using the operatorial formalism described below.

Operators

According to its mathematical definition an *operator* transforms one function into another function. The operation 'differentiate with respect to x' carried out on the function $f(x) = x^2$ transforms it into a new function, $g(x) = 2x$. In other words,

$$g(x) = \hat{\Omega} f(x), \tag{1.16}$$

where $\hat{\Omega}$ is the operator $\equiv \frac{d}{dx}$. In quantum mechanics all observable properties such as position, momentum, angular momentum and energy are represented by a particular operator, examples of which are listed in Table 1.1. Supposing we know the wavefunction describing the system at a particular time (e.g., the motion of the electron in an atom determined by solving the Schrödinger equation), then we can use the operators to determine what the corresponding observables are. First we need to determine whether the wavefunction is an *eigenfunction* of the particular operator $\hat{\Omega}$ as defined by the relationship;

$$\hat{\Omega}\Psi = \omega\Psi. \tag{1.17}$$

If applying $\hat{\Omega}$ to the wavefunction gives ω as a *constant* independent of position coordinates, then Ψ is called an *eigenfunction* of the operator and ω is the *eigenvalue.* If the relationship is obeyed then the eigenvalue is *equal* to the value of the observable property, and that observable is precisely determined;

Table 1.1

Quantum mechanical operators representing observables

Observable	Operator	
Position	$\hat{x}$	$x\times$ (multiply by x)
Momentum (component)	$\hat{p}_x$	$-i\hbar\frac{\partial}{\partial x}$
Angular momentum (cpt)	$\hat{l}_x$	$-i\hbar(y\frac{\partial}{\partial z} - z\frac{\partial}{\partial y})$
Kinetic energy	$\hat{T}$	$\frac{-\hbar^2}{2m}[\frac{\partial^2}{\partial x^2} + \frac{\partial^2}{\partial y^2} + \frac{\partial^2}{\partial z^2}]$
Potential energy	$\hat{V}$	$V(x, y, z)$

i.e., the result of every measurement we would make of that observable would be exactly the same value ω. If, on the other hand, the relationship (1.17) is not obeyed, then the observable cannot be precisely determined; we can only define the mean value of that observable using the *expectation value* $\langle\omega\rangle$,

$$\langle\omega\rangle = \frac{\int \Psi^*\hat{\Omega}\Psi d\tau}{\int \Psi^*\Psi d\tau}. \tag{1.18}$$

A number of measurements of the observable would give a different result each time with an average value equal to the expectation value.

The Hamiltonian operator

The total energy of a particle is always a well-defined quantity, provided that the system is not actively undergoing a change (as might occur in the collision of one atom with another); therefore the wavefunction describing such a *stationary state* of the system must obey an eigenvalue equation for the total energy operator. The total energy is always a sum of potential and kinetic energies, and the operator representing the total energy is called the Hamiltonian operator $\mathcal{H}$ (note that the potential and kinetic energies are not conserved quantities individually). The wavefunction must obey the equation

$$\mathcal{H}\Psi = E\Psi, \tag{1.19}$$

where $\mathcal{H} = \hat{T} + \hat{V}$ is the sum of the kinetic energy operator $\hat{T}$, and the potential energy operator $\hat{V}$. Using the operators given in Table 1.1, we obtain the Schrödinger equation for the motion of a single particle,

$$(\hat{T}+\hat{V})\Psi \equiv \left[\frac{-\hbar^2}{2m}\nabla^2 + V(x,y,z)\right]\Psi = E\Psi. \qquad (1.20)$$

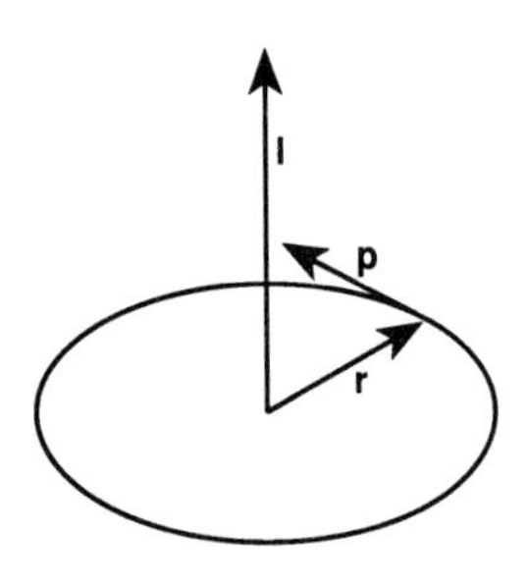

Fig. 1.11 The classical definition of angular momentum, which is represented by a vector **l** directed perpendicular to the plane of rotation.

1.5 Angular momentum

Classical angular momentum

Classically the angular momentum of a particle, **l** is defined by the equation

$$\mathbf{l} = \mathbf{r} \times \mathbf{p}, \qquad (1.21)$$

where **r** is the position vector and **p** the linear momentum vector (see Fig. 1.11). For a system of particles the total angular momentum **L** is

$$\mathbf{L} = \sum_i \mathbf{r}_i \times \mathbf{p}_i. \qquad (1.22)$$

The angular momentum is directly related to the force (or more correctly the 'torque') that would be needed to stop the body rotating. For a rigid body rotating about a fixed axis, the equation reduces to

$$L = I\omega, \qquad (1.23)$$

where I is the moment of inertia, $I = \sum_i m_i r_i^2$; the summation is over all particles, mass m_i making up the body, r_i is the distance of a particle to the fixed axis, and ω is the angular velocity (radians s^{-1}). Strictly speaking, angular momentum is a *vector* quantity, as defined in eqn (1.21); eqn (1.23) just gives the magnitude of that vector. The direction of the vector is parallel to the direction of the rotation axis, and by convention, an observer looking along the axis towards the angular momentum vector will see anticlockwise rotation.

An electron orbiting an atom does not confine itself to rotation about a single axis, but it follows a trajectory such that the rotation axis varies with time. To deal with this situation we need to define a fixed axis system x, y, z, and then at any instant can specify the components of the angular momentum vector ℓ_x, ℓ_y, ℓ_z along these three mutually perpendicular axes. The *total angular momentum* is then defined as the vector **l** where

$$\mathbf{l} = \ell_x\mathbf{i} + \ell_y\mathbf{j} + \ell_z\mathbf{k}, \qquad (1.24)$$

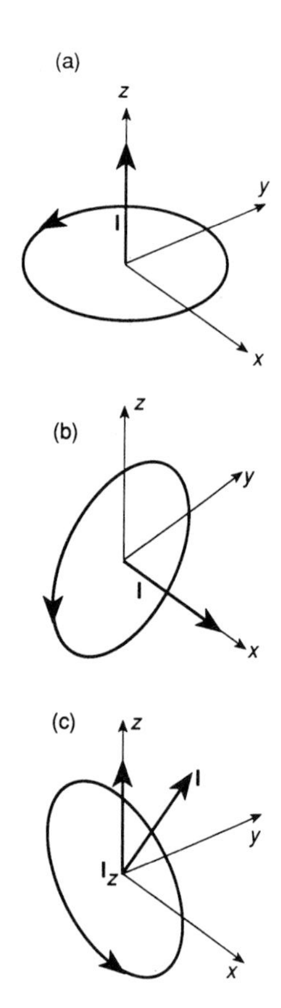

Fig. 1.12 The angular momentum vectors for:
(a) $\ell_x = \ell_y = 0, \ell_z = |\mathbf{l}|$,
(b) $\ell_y = \ell_z = 0, \ell_x = |\mathbf{l}|$,
(c) $\ell_x \neq \ell_y \neq \ell_z \neq 0$.

i, **j**, **k** being unit vectors along x, y, and z respectively. Thus, to give a macroscopic illustration, an Earth-orbiting satellite following the equator would have $\ell_x = \ell_y = 0, \ell_z = |\mathbf{l}|$ while if it followed the Greenwich Meridian it would have $\ell_y = \ell_z = 0, \ell_x = |\mathbf{l}|$ (see Fig. 1.12). Clearly there are other circular paths for which all three components might be nonzero (Fig. 1.12c). Using standard rules for vector multiplication we obtain from eqn (1.21),

$$\ell_x = yp_z - zp_y \qquad \ell_y = zp_x - xp_z \qquad \ell_z = xp_y - yp_x. \qquad (1.25)$$

The magnitude of the total angular momentum is

$$\ell = \sqrt{\ell_x^2 + \ell_y^2 + \ell_z^2}. \tag{1.26}$$

Quantum mechanical angular momentum

In quantum mechanics we obtain operators for the components of the angular momentum by replacing x and p_x, etc., by their corresponding operators given in Table 1.1, e.g.,

$$\hat{\ell}_z = \frac{\hbar}{i}\left[x\frac{\partial}{\partial y} - y\frac{\partial}{\partial x}\right]. \tag{1.27}$$

We cannot define an operator for ℓ itself, because we cannot take the square root of an operator as is implied by eqn (1.26). However the operator for ℓ^2 can be defined as

$$\hat{\ell^2} = \hat{\ell}_x\hat{\ell}_x + \hat{\ell}_y\hat{\ell}_y + \hat{\ell}_z\hat{\ell}_z. \tag{1.28}$$

More conveniently, however, the operators for ℓ^2 and ℓ_z can be written in spherical polar coordinates (see Fig. 1.13) as,

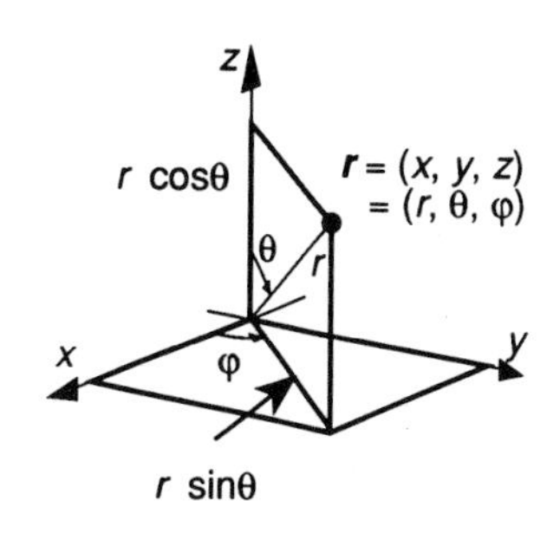

Fig. 1.13 Spherical polar coordinates: r is the distance from the position of the particle to the origin, θ is the angle with respect to the z-axis, and ϕ is the angle with respect to the x-axis of the projection of the position vector $\mathbf{r}$ onto the xy plane. The transformation to Cartesian coordinates is $x = r\sin\theta\cos\phi$, $y = r\sin\theta\sin\phi$, $z = r\cos\theta$.

$$\hat{\ell}_z = \frac{\hbar}{i}\frac{\partial}{\partial\phi}, \tag{1.29}$$

$$\hat{\ell^2} = -\Lambda^2\hbar^2,$$

$$\Lambda^2 = \frac{1}{\sin^2\theta}\frac{\partial^2}{\partial\phi^2} + \frac{1}{\sin\theta}\frac{\partial}{\partial\theta}\sin\theta\frac{\partial}{\partial\theta}. \tag{1.30}$$

Commutation of angular momentum operators

A *commutator* is a special type of operator defined as

$$[A, B] = \hat{A}\hat{B} - \hat{B}\hat{A}. \tag{1.31}$$

$[A, B]$ is the commutator and $\hat{A}$ and $\hat{B}$ are two operators. Note that the product of two operators $\hat{A}\hat{B}$ is to be interpreted as, first operate with $\hat{B}$ on the function, then operate on the result with $\hat{A}$. It can be shown quite easily that if two operators are *commutative*, implying $[A, B] = 0$, then any function which is an eigenfunction of $\hat{A}$ is simultaneously an eigenfunction of $\hat{B}$. If these operators correspond to real observables as those in Table 1.1, then the implication is that it is possible to specify both observables simultaneously with arbitrary precision. The Heisenberg uncertainty principle is an expression of the fact that the position and momentum operators do *not* commute, and therefore these two observables cannot both be specified precisely.

We note here the following very important relationships that are derived in standard quantum mechanics textbooks.

$$[\ell^2, \ell_z] = [\ell^2, \ell_x] = [\ell^2, \ell_y] = 0, \tag{1.32}$$

$$[\ell_x, \ell_y] = i\hbar\hat{\ell}_z \qquad [\ell_y, \ell_z] = i\hbar\hat{\ell}_x \qquad [\ell_z, \ell_x] = i\hbar\hat{\ell}_y. \tag{1.33}$$

Furthermore for an electron in an atom subject to an isotropic electrostatic potential, $\hat{\ell^2}$ and its components are commutative with the Hamiltonian operator

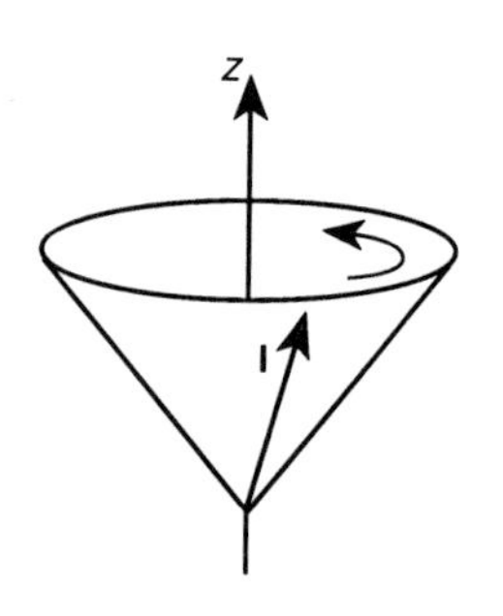

Fig. 1.14 Cone of precession of the angular momentum vector, with well-defined z-component.

$$[\ell^2, \mathcal{H}] = 0 \qquad [l_z, \mathcal{H}] = 0. \tag{1.34}$$

Therefore any wavefunction which is an eigenfunction of the energy operator (Hamiltonian) will simultaneously be an eigenfunction of $\hat{\ell}^2$ and $\hat{\ell}_z$.

It is a remarkable consequence of the uncertainty intrinsic to quantum mechanics that, whereas it is possible to define operators for the various components, ℓ_x, ℓ_y, ℓ_z, it is only possible for the wavefunction to be an eigenfunction of *one* of these operators, as implied by the non-commutation in (1.33). Normally the wavefunction is chosen so that it has a well-defined z-component in addition to the well-defined value of ℓ^2. As the x- and y-components are undefined the quantum mechanical angular momentum vector is often represented as a cone centred on the z-axis as shown in Fig. 1.14, in which the direction of $\mathbf{l}$ precesses around the axis.

Energy levels for a particle on a sphere

The quantum mechanical description of a particle confined to motion on the surface of a sphere of radius R is of great value in solving the Schrödinger equation for atoms, and therefore it is considered here in detail. The Schrödinger equation for this hypothetical situation is most conveniently cast in spherical polar coordinates,

$$\frac{-\hbar^2}{2m}\left[\frac{1}{r}\frac{\partial^2}{\partial r^2}r + \frac{1}{r^2}\Lambda^2\right]\Psi + V(r,\theta,\phi)\Psi = E\Psi, \tag{1.35}$$

where the Legendrian operator Λ^2 is given in eqn (1.30). For a particle confined to a sphere, $V(r,\theta,\phi) = 0$ for $r = R$ and $V(r,\theta,\phi) = \infty$ for $r \neq R$. The wavefunction must be zero except at $r = R$ and therefore we can ignore the r-dependent derivative terms in the kinetic energy operator, reducing the Schrödinger equation to a second-order differential equation in two variables, θ and ϕ,

$$\frac{-\hbar^2}{2mR^2}\Lambda^2\Psi = E\Psi. \tag{1.36}$$

We make a trial separation of variables of the form

$$\Psi(\theta,\phi) = \Theta(\theta)\Phi(\phi). \tag{1.37}$$

Substituting (1.37) into (1.36) and treating θ as a constant gives an equation for $\Phi(\phi)$,

$$\frac{-\hbar^2}{2mR^2}\frac{\partial^2}{\partial\phi^2}\Phi(\phi) = (const.)\Phi(\phi). \tag{1.38}$$

Note that this is precisely the form of the Schrödinger equation for a particle confined to motion on a circular ring. A general solution is

$$\Phi(\phi) = A\exp(iM\phi) + B\exp(-iM\phi), \tag{1.39}$$

where A, B and M are arbitrary constants. Substituting (1.39) into (1.38) gives

$$\frac{-\hbar^2}{2mR^2}\frac{\partial^2}{\partial\phi^2}\Phi(\phi) = \frac{\hbar^2 M^2}{2mR^2}\Phi(\phi). \tag{1.40}$$

However, in order to be consistent with the Born interpretation (Section 1.4) we require that the wavefunction must be single valued. This leads to a *boundary condition* imposed on the wavefunction, that

$$\Phi(\phi) = \Phi(\phi + 2n\pi), \tag{1.41}$$

where n is any integer, because the angles ϕ and $\phi + 2n\pi$ are equivalent in real space. In order to satisfy the boundary conditions (1.41), only those solutions (1.39) which have M is integral are valid, and therefore M is a *quantum number* because it labels and characterizes the permitted solutions of the equation. As the constants A and B are arbitrary in (1.39), it is common practice to choose $A = 1/\sqrt{2\pi}$ and $B = 0$, thus giving a normalized set of solutions which are eigenfunctions of the ℓ^2 and ℓ_z operators (see eqns (1.46) and (1.47)).

The solution of the θ-dependent equation is obtained by inserting eqn (1.40) into eqn (1.36) and applying a boundary condition analogous to (1.41);

$$\Theta(\theta) = \Theta(\theta + 2n\pi), \tag{1.42}$$

$$\frac{\hbar^2}{2mR^2}\left[\frac{M^2}{\sin^2\theta} - \frac{1}{\sin\theta}\frac{\partial}{\partial\theta}\sin\theta\frac{\partial}{\partial\theta}\right]\Theta(\theta) = E\Theta(\theta). \tag{1.43}$$

The solutions of this equation, obtained by standard mathematical techniques, are known as the associated Legendre functions. The quantum number M must also characterize these solutions as it appears explicitly in eqn (1.43), but in addition, there is a second quantum number l, which arises because of the boundary condition (1.42). The permitted values of l cannot be less than M. The product of Θ and Φ is known as a *spherical harmonic function* and is given the symbol $Y_{lM}(\theta, \phi)$. The first few spherical harmonic functions are listed in Table 1.2. It can be shown by insertion that

$$\Lambda^2 Y_{lM} = -l(l+1)Y_{lM}, \tag{1.44}$$

and consequently from (1.36) the energy of a particle confined to a sphere is

Table 1.2

Spherical harmonic functions

l	M	$Y_{lM}(\theta, \phi)$
0	0	$1/(2\pi^{\frac{1}{2}})$
1	0	$\frac{1}{2}(\frac{3}{\pi})^{\frac{1}{2}}\cos\theta$
	±1	$\mp\frac{1}{2}(\frac{3}{2\pi})^{\frac{1}{2}}\sin\theta e^{\pm i\phi}$
2	0	$\frac{1}{4}(\frac{5}{\pi})^{\frac{1}{2}}(3\cos^2\theta - 1)$
2	±1	$\mp\frac{1}{2}(\frac{15}{2\pi})^{\frac{1}{2}}\cos\theta\sin\theta e^{\pm i\phi}$
2	±2	$\frac{1}{4}(\frac{15}{2\pi})^{\frac{1}{2}}\sin^2\theta e^{\pm 2i\phi}$

$$E = \frac{\hbar^2}{2mR^2} l(l+1). \tag{1.45}$$

The permitted values of the energy are therefore *quantized* because l must be a positive integer.

Having obtained the wavefunctions we can extract other information about the particle: the operator for the square of the orbital angular momentum $\hat{l}^2$ is just $-\Lambda^2\hbar^2$ and therefore

$$\hat{l}^2 Y_{lM} = l(l+1)\hbar^2 Y_{lM}. \tag{1.46}$$

The operator for the z-component of the angular momentum is $\hat{l}_z = -i\hbar\frac{\partial}{\partial\phi}$. The spherical harmonic functions are eigenfunctions of this operator with eigenvalue $M\hbar$

$$\hat{l}_z Y_{lM} = M\hbar Y_{lM}. \tag{1.47}$$

In other words, the wavefunction defined by quantum numbers l and M represents a particle on a sphere with total angular momentum equal to $[l(l+1)]^{\frac{1}{2}}\hbar$, z-component equal to $M\hbar$, and energy $\frac{\hbar^2}{2mR^2}l(l+1)$.

2 The structure and spectrum of the hydrogen atom

2.1 Practical emission spectroscopy

The fraction of hydrogen in the atomic form in a room temperature sample is negligibly small under normal conditions, as molecular hydrogen, H_2, is the standard form. The emission spectrum of atomic hydrogen is therefore normally recorded by setting up an electric discharge in a low pressure hydrogen lamp. An illustration of a suitable tube is given in Fig. 2.1. The chemical and physical processes taking place in the discharge are complex and strongly dependent on conditions such as the pressure and voltage applied. The breakdown of the gas is a result of a few ionized particles, present originally in the gas sample, being accelerated to sufficient energy to ionize further atoms and molecules by high energy collisions, setting off an avalanche of ionization events. The collisions can also cause breaking of bonds and hence the production of hydrogen atoms. Atomic hydrogen may alternatively be formed via the recombination of slow electrons with protons.

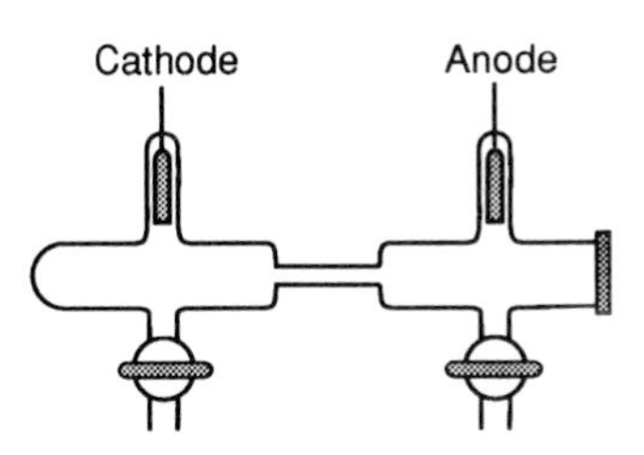

Fig. 2.1 A discharge tube for recording the emission spectrum of the hydrogen atom. A typical pressure of the order of 1 mbar H_2 is used. The central capillary helps to spatially confine the discharge, enhancing current density and brightness.

The hydrogen atoms are typically formed in a wide distribution of different quantum states and therefore emission back to lower levels is possible. The effective temperature of the atoms in the discharge is many thousands of Kelvin. The light emitted by the discharge can be focused onto the entrance slit of a photographic spectrograph of the type shown in Fig. 1.1. An alternative configuration is to monitor the light passing through an exit slit of the monochromator, as shown in Fig. 2.2, as the prism (or alternatively a diffraction grating) is rotated. The rotation has the effect of slowly changing the wavelength of light

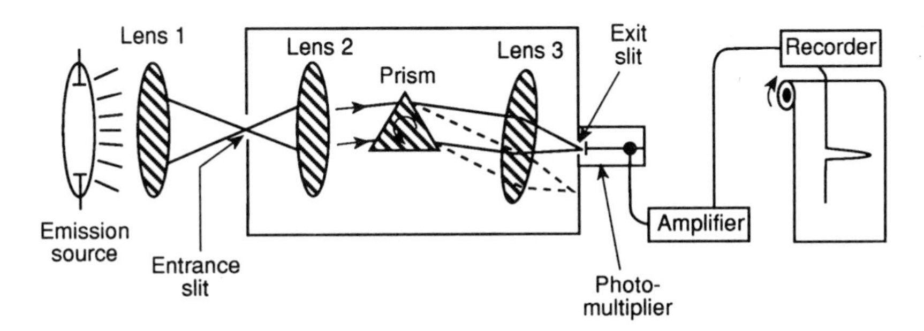

Fig. 2.2 Schematic diagram of a spectrometer employing a photomultiplier as detector. Light from the discharge is focused onto the entrance slit and is then dispersed by a prism. Only a very narrow band of wavelengths passes through the exit slit and the intensity of light is monitored by the photomultiplier. The prism is continuously rotated, scanning the detected wavelength.

transmitted though the exit slit and therefore the intensity can be plotted as a function of wavelength. A photomultiplier tube is commonly used for detecting the light; in this device the light incident on the front electrode produces some electrons via the photoelectric effect, which are then accelerated onto a chain of subsequent electrodes, producing ever-increasing numbers of secondary electrons. A single photon can be converted to as many as 10^8 electrons and is then detected as an electrical pulse. Individual photons can be counted if necessary, leaving the experimentalist in no doubt that the photon description of light is a reality!

The emission spectrum of hydrogen shows the sharp line structure associated with transitions between its energy states as discussed in Section 1.1. Many series of lines have been observed, some of which are illustrated in Fig. 2.3. The Lyman series occurs in the vacuum ultraviolet region of the spectrum and is rather more difficult to obtain in practice than suggested by the simple set-up in Fig. 2.2. The emitted light would be absorbed by air molecules and

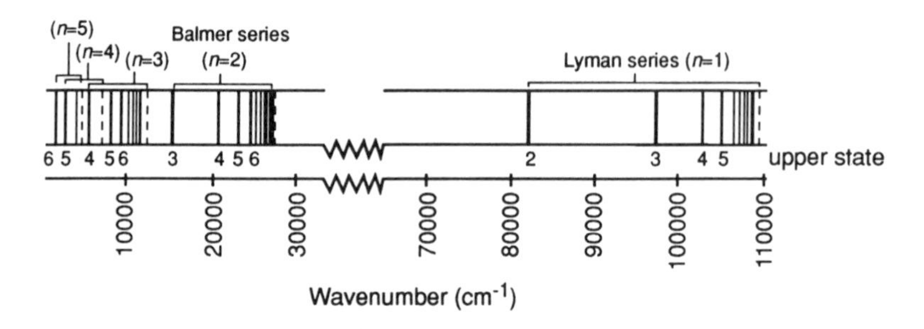

Fig. 2.3 Schematic representation of the H-atom emission spectrum, ranging from the infrared (left side) to the vacuum ultraviolet (right side). The thickness of lines indicates the approximate intensity, and the dotted lines correspond to series limits. The lower state principal quantum numbers are given above the spectrum. The $n = 3$, $n = 4$, and $n = 5$ sets are known as the Paschen, Brackett, and Pfund series respectively.

therefore it would be necessary to use a vacuum spectrometer, in which the light traverses a completely evacuated path.

In 1890, Rydberg showed that the frequencies of the observed emission lines can all be fitted to a single formula,

$$\nu = -c\mathcal{R}\left(\frac{1}{n_1^2} - \frac{1}{n_2^2}\right), \tag{2.1}$$

where $n_2 = 1, 2, 3, 4$ for the Lyman, Balmer, Paschen and Brackett series respectively and n_1 is an integer greater than n_2. $\mathcal{R}$ is known as the Rydberg constant, and c is the velocity of light. Bohr successfully explained the observed frequencies by requiring the electron orbiting the nucleus to be confined to a restricted set of elliptical orbits. The total energy of the electron in these orbits was calculated classically to be

$$E = \frac{-hc\mathcal{R}}{n^2} \qquad \mathcal{R} = \frac{\mu e^4}{8\epsilon_0^2 h^3 c}. \tag{2.2}$$

In this equation μ is the reduced mass of the electron,

$$\mu = \frac{m_e . m_N}{m_e + m_N}, \tag{2.3}$$

which is almost equal to the electron mass m_e because the nuclear mass $m_N \gg m_e$. The Rydberg formula, eqn (2.1), then arises from eqn (2.2) and the equation $\Delta E = h\nu$ by taking the differences of the energy levels with different n values, where n is an integer. However, Bohr's theory provided no explanation for the restriction of n to integral values. The significance of this quantum number becomes apparent in terms of Schrödinger's wave mechanical view of the atom.

2.2 The Schrödinger equation for the hydrogen atom

The wavefunctions Ψ and energy levels, E, for the hydrogen atom can be obtained by solving the Schrödinger equation, which may be written in this case as,

$$\left[-\frac{\hbar^2}{2\mu}\nabla^2 + V(r)\right]\Psi = E\Psi. \tag{2.4}$$

The reduced mass occurs instead of the true electron mass (*cf.* eqn 1.11) because we have referred the coordinate system to an origin fixed at the nucleus (and moving with it) rather than an arbitrary laboratory-fixed origin. Note that there is a slight dependence of μ on the nuclear mass in eqn (2.3), a point we shall return to later. The operator ∇^2 may be written either in the Cartesian coordinate form,

$$\nabla^2 = \frac{\partial^2}{\partial x^2} + \frac{\partial^2}{\partial y^2} + \frac{\partial^2}{\partial z^2}, \tag{2.5}$$

or more usefully in spherical polar coordinates (see Fig. 1.13)

$$\nabla^2 = \frac{1}{r}\frac{\partial^2}{\partial r^2}r + \frac{1}{r^2}\Lambda^2, \tag{2.6}$$

where

$$\Lambda^2 = \frac{1}{\sin^2\theta}\frac{\partial^2}{\partial\phi^2} + \frac{1}{\sin\theta}\frac{\partial}{\partial\theta}\sin\theta\frac{\partial}{\partial\theta}. \tag{2.7}$$

The first term in the square brackets in (2.4) is the kinetic energy operator $\hat{T}$ while the second term is the potential energy operator $\hat{V}$; for the hydrogen atom $\hat{V}$ is simply determined by the electron–nucleus attraction given by

$$V(r) = \frac{-Ze^2}{4\pi\epsilon_0 r}, \tag{2.8}$$

where Z is the atomic number, 1 for hydrogen.

Boundary conditions

The Born interpretation (Section 1.4) gives a physical meaning to the wavefunction, and in solving the Schrödinger equation we must impose *boundary conditions* that ensure the physical interpretation is possible. In the case of the hydrogen atom, we can first recognize that the electron must exist at a finite distance from the nucleus and therefore require that

$$\Psi \to 0 \text{ as } r \to \infty. \tag{2.9}$$

Ψ must also have finite (or zero) values for all r, including the origin. Secondly, we note (as in the case of the wavefunctions for a particle confined to a sphere, Section 1.5) that the angles θ and ϕ are equivalent in physical terms to the angles $\theta + 2\pi$ and $\phi + 2\pi$ respectively, and therefore we require

$$\Psi(r, \theta, \phi) = \Psi(r, [\theta + 2n\pi], [\phi + 2m\pi]), \quad (2.10)$$

where n and m are any pair of integers. The consequence of these boundary conditions is that physically meaningful solutions of the Schrödinger equation are not obtainable at any energy E, but as shown in Section 2.5, only at specific quantized energies

$$E = -\frac{\mu e^4}{32\pi^2\epsilon_0^2\hbar^2 n^2}, \quad (2.11)$$

where n is permitted the values 1,2,3... The calculated energy is exactly equal to that obtained in the Bohr theory of the atom, but now an explanation is offered for the restriction on values of n; the permitted energies correspond to standing-wave solutions of the wave equation, and n is related to the number of *nodes* in the wavefunction (see Section 2.4).

Table 2.1

Hydrogenic radial wavefunctions

n	l	$R_{nl}(r)$
1	0	$(Z/a_0)^{\frac{3}{2}} 2\exp(-\rho/2)$
2	0	$(Z/a_0)^{\frac{3}{2}}(1/2\sqrt{2})(2-\rho)\exp(-\rho/2)$
2	1	$(Z/a_0)^{\frac{3}{2}}(1/2\sqrt{6})\rho\exp(-\rho/2)$
3	0	$(Z/a_0)^{\frac{3}{2}}(1/9\sqrt{3})(6-6\rho+\rho^2)\exp(-\rho/2)$
3	1	$(Z/a_0)^{\frac{3}{2}}(1/9\sqrt{6})(4-\rho)\rho\exp(-\rho/2)$
3	2	$(Z/a_0)^{\frac{3}{2}}(1/9\sqrt{30})\rho^2\exp(-\rho/2)$

where $\rho = 2Zr/(na_0)$, $a_0 = 0.05292$ nm

2.3 The wavefunctions

As will be shown in Section 2.5 the wavefunctions, obtained by solving the Schrödinger equation are written as a product of a radial function $R_{nl}(r)$ and an angular function $Y_{lm_l}(\theta, \phi)$

$$\Psi_{nlm_l} = R_{nl}(r)Y_{lm_l}(\theta, \phi). \quad (2.12)$$

The radial wavefunction is an associated Laguerre function, labelled by two quantum numbers n and l, as listed in Table 2.1. The angular wavefunction is a spherical harmonic function (Table 1.2) specified by the two quantum numbers

l and m_l. The angular wavefunctions are identical to those obtained in Section 1.5 for a particle confined to the surface of a sphere. The total set of allowed wavefunctions is obtained by combining each radial wavefunction with the set of $2l+1$ angular wavefunctions having the same label l, but different m_l values. Each wavefunction, defined by the set of quantum numbers n, l, and m_l must have an associated energy as defined by eqn (2.11). Note, however, that as that equation only depends on the label n, there is, for all values of n except 1, more than one total wavefunction corresponding to each energy level. This phenomenon is crucially important in spectroscopy and quantum mechanics and is known as *degeneracy*. The key point here is that the electron can exist in more than one physically distinct state in the atom without changing its total energy.

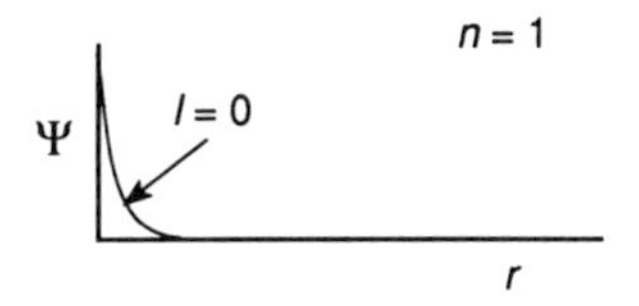

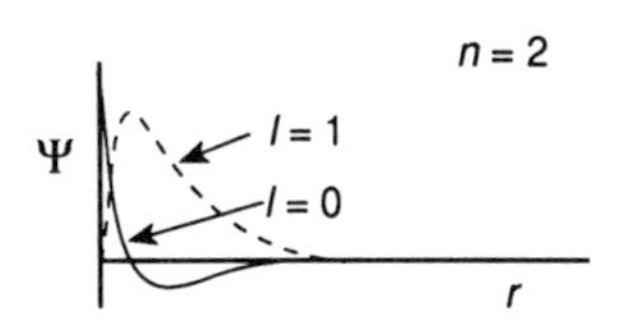

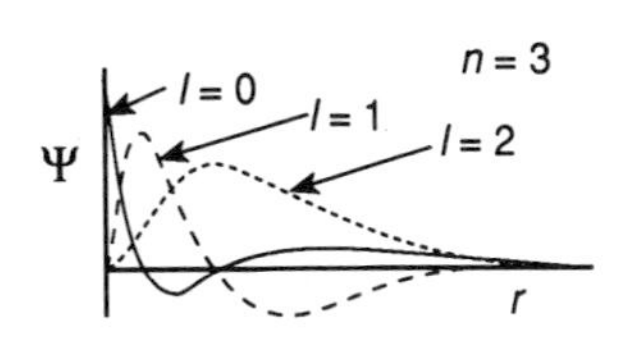

Fig. 2.4 The H-atom radial wavefunctions for $n = 1$ to 3.

Orbitals

The wavefunctions for the atom are generally referred to as *orbitals*. This word represents an extension of the old ideas of quantum theory (Bohr theory) in which electrons moved in well-defined elliptical orbits around the nucleus. In the modern quantum theory the orbital trajectory is not precisely specified; the true meaning of the orbitals is seen through the Born interpretation of the wavefunctions, namely that the square of the wavefunction gives the probability density for the electron at a particular point in space (Section 1.4, eqn 1.14).

The radial parts of the wavefunction corresponding to the functions given in Table 2.1 are plotted in Fig. 2.4 as a function of r for $n = 1, 2$, and 3, while in Fig. 2.5 the probability density functions Ψ^2 are plotted. (Note that the radial wavefunctions are real so that $\Psi^*\Psi = \Psi^2$.) Some remarkable facts emerge; first, for an $l = 0$ orbital the greatest probability density Ψ^2 occurs at $r = 0$, suggesting that the electron has a significant probability of being found *inside* the nucleus. There is strong experimental evidence to support this prediction; see Section 5.7 for a discussion of the Fermi–contact interaction. Secondly, an electron in an $l = 1$ or 2 orbital has no probability of being found at the nucleus. This is a consequence of the nonzero angular momentum of the electron in these states (see Section 2.4); the electron is flung away from the nucleus by centrifugal forces. Thirdly, for the states with $n - l > 1$, there are radii other than $r = 0$ or $r = \infty$ with zero probability density. These positions are called *radial nodes*; it is an interesting effect that an electron can apparently exist on one or other side of these positions but can never exist at the node.

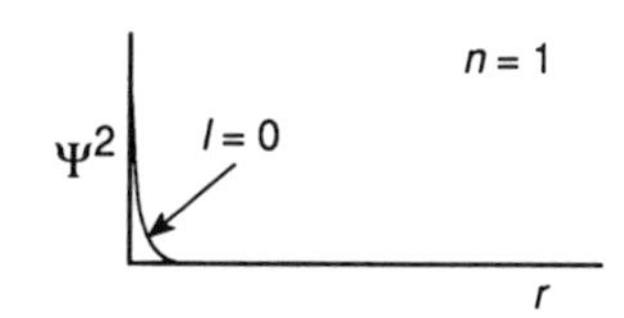

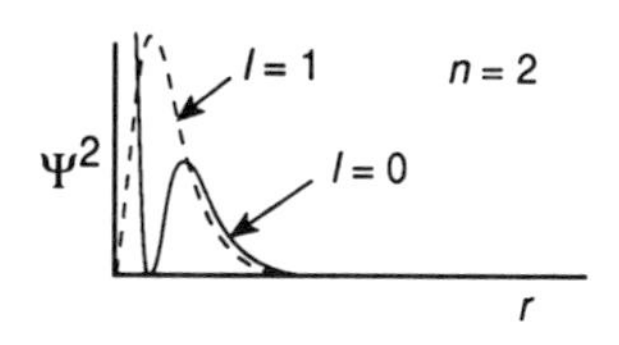

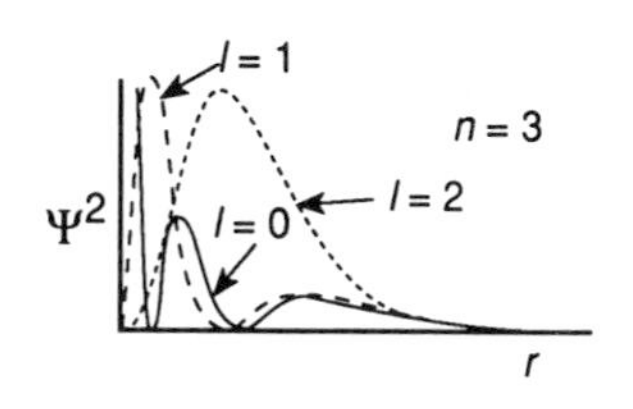

In order to gain a better understanding of the orbitals it is convenient to define the *radial distribution function* $P(r)$, such that the probability of finding an electron in a spherical shell of radius r and thickness dr is $P(r)\mathrm{d}r$.

$$P(r)\mathrm{d}r = \int_0^{2\pi} \mathrm{d}\phi \int_0^{\pi} \mathrm{d}\theta \Psi^*\Psi r^2 \sin\theta \mathrm{d}r. \tag{2.13}$$

Fig. 2.5 The probability density functions for $n = 1$ to 3, defined such that the probability of finding the electron in the volume element $\mathrm{d}\tau$ at position (r, θ, ϕ) is $\Psi^*\Psi(r, \theta, \phi)\,\mathrm{d}\tau$.

(In spherical polar coordinates the volume element $\mathrm{d}\tau = r^2\sin\theta\, \mathrm{d}r\mathrm{d}\theta\mathrm{d}\phi$.) For a spherically symmetric orbital, which has no dependence on θ, or ϕ, the equation becomes

$$P(r)\mathrm{d}r = 4\pi r^2 \Psi^*\Psi \mathrm{d}r. \tag{2.14}$$

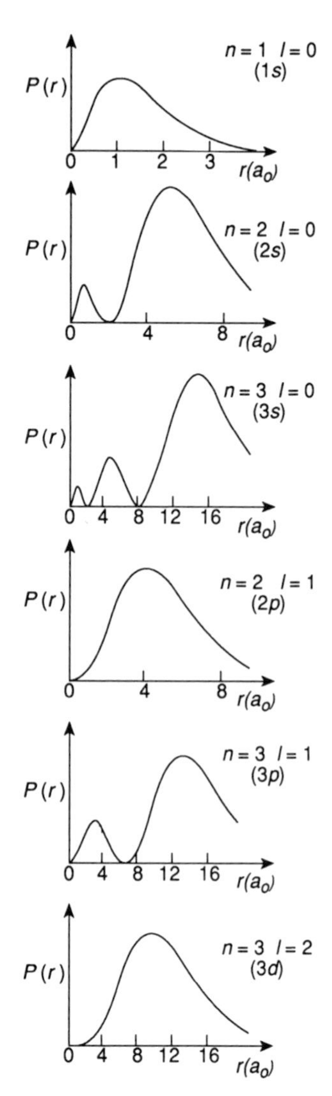

Fig. 2.6 The H-atom radial distribution functions, defined in eqn (2.13) for $n = 1$ to 3. The horizontal scale is in units of Bohr radii ($1a_0 = 0.05292$ nm).

The difference between the radial distribution function and the probability density is that the latter is the probability of finding the electron at a particular point (r, θ, ϕ) in a volume element $\mathrm{d}\tau$, whereas the former is the probability of finding an electron at a distance r from the nucleus integrated over all angles. Figure 2.6 plots the radial distribution function for the $n = 1, 2$ and 3, $l = 0, 1$ and 2 orbitals. It can be seen that with inclusion of the r^2 factor the probability of finding an electron at $r = 0$ is now zero, for all the functions. However, comparing the distributions for $n = 2$, $l = 0$ and 1 shows that there is greater chance of finding the $l = 0$ electron very close to the nucleus than the $l = 1$ electron. Nevertheless, the $l = 0$ distribution also has a large maximum beyond the maximum of the $l = 1$ distribution and the counterbalancing effect is such that the mean potential energy of interaction with the nucleus for an electron in the two types of orbital is actually equal: the mean kinetic energies are also equal and therefore these states are degenerate.

There is a difficulty in representing the *angular* parts of the H-atom wavefunctions (the spherical harmonics) in a polar plot because, as can be seen in Table 1.2, these are *complex* functions. However, an important result of quantum mechanics states that any linear combination of degenerate wavefunctions, which are solutions of the Schrödinger equation, must also be a valid solution of the Schrödinger equation. It is therefore common to represent the angular wavefunctions by *real* linear combinations. For the $l = 1$ set (p-orbitals) these combinations are:

$$\begin{aligned}\Psi(p_x) &= \tfrac{1}{\sqrt{2}}(Y_{11} + Y_{1-1}),\\ \Psi(p_y) &= \tfrac{i}{\sqrt{2}}(Y_{11} - Y_{1-1}),\\ \Psi(p_z) &= Y_{10}.\end{aligned} \tag{2.15}$$

Sketches of the angular parts of the H-atom wavefunction in their real forms are shown in a polar plot in Fig. 2.7. It should be noted that for some of the functions there are *angular nodes*, meaning that the probability is zero for finding the electron at a particular angle with respect to the origin. An alternative representation that looks very similar is to draw a probability surface as in Fig. 2.8; these boundary surfaces represent the volume within which there is a 90% chance of finding the electron, and the dimensions are drawn in a particular direction so as to show the relative probability for that direction.

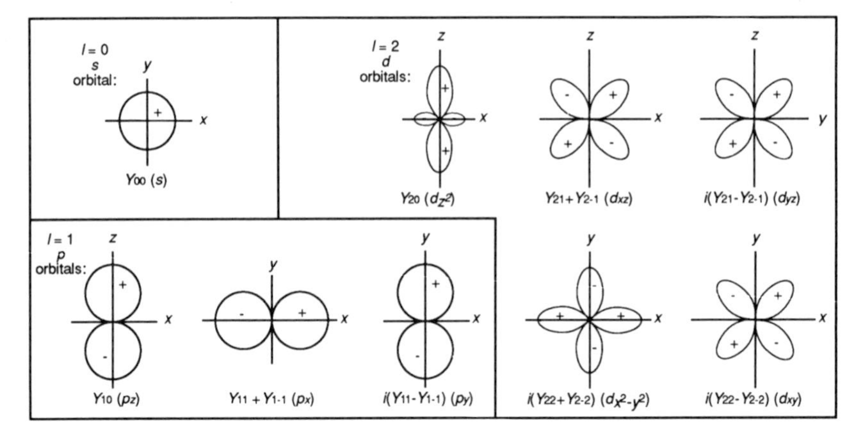

Fig. 2.7 Polar diagrams of the real combinations of spherical harmonics for $l = 0$, 1 and 2. The length of a line from the origin to the boundary gives the magnitude of the wavefunction in that direction.

2.4 The quantum numbers

The principal quantum number n

The principal quantum number n is permitted the integer values 1,2,3 ... and its significance is that it determines the energy

$$E = -\frac{\mathcal{R}hc}{n^2}, \tag{2.16}$$

where $\mathcal{R}$ is the collection of constants known as the Rydberg constant

$$\mathcal{R} = \frac{\mu e^4}{8\epsilon_0^2 h^3 c}. \tag{2.17}$$

It is normally expressed in the non-SI units $= 109\,678\ \mathrm{cm}^{-1}$. The energy does not depend on any of the other quantum numbers. The energy levels are depicted in Fig. 2.9; these take the discrete values $-\mathcal{R}hc, -\frac{1}{4}\mathcal{R}hc, -\frac{1}{9}\mathcal{R}hc \ldots$. The lowest energy level is the $n = 1$ state and is referred to as the *ground state*.

The mean radius of the electron from the nucleus shows a strong dependence on the principal quantum number, as can be seen in Figs 2.4 to 2.6; it can be shown that for an $l = 0$ radial function the mean radius is given by

$$\langle r \rangle = \frac{3}{2}\frac{a_0 n^2}{Z} \qquad a_0 = \frac{4\pi\epsilon_0\hbar^2}{\mu e^2}. \tag{2.18}$$

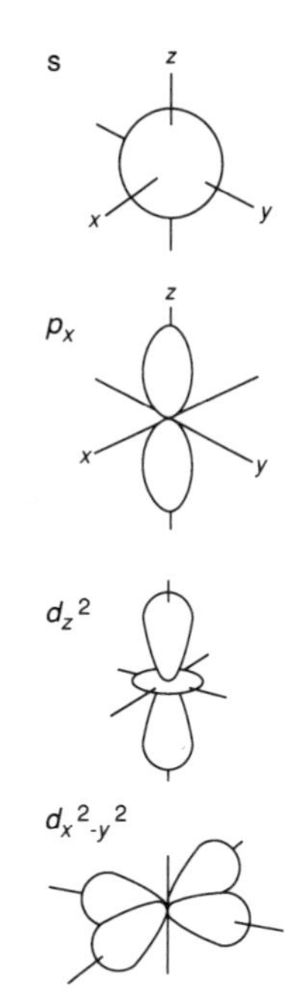

Fig. 2.8 Boundary surfaces for the $l = 0$, 1 and 2 orbitals, showing the volume within which 90% of the probability distribution is located.

The orbital angular momentum quantum number l

The quantum number l is directly related to the magnitude of the orbital angular momentum of the electron $|\,\mathbf{l}\,|$ through eqn (2.19),

$$|\,\mathbf{l}\,| = [l(l+1)]^{\frac{1}{2}}\hbar. \tag{2.19}$$

The spherical harmonics are eigenfunctions of the $\hat{\ell}^2$ operator as shown in Section 1.5, with eigenvalue $l(l+1)\hbar^2$. l is permitted to take integral values from 0 to $n-1$. States with the same n but different l have different radial wavefunctions; the number of radial nodes in the wavefunction is always equal to $n - l - 1$ (see Fig. 2.4) and l is equal to the number of angular nodal planes in the wavefunction (Fig. 2.7). States with $l = 0, 1, 2, 3, 4, 5 \ldots$ are labelled for historical reasons as *s, p, d, f, g, h* ... respectively and the notation 1*s*, 2*p*, 3*p*, 3*d*, etc., is used for the orbitals, the first number being the principal quantum number.

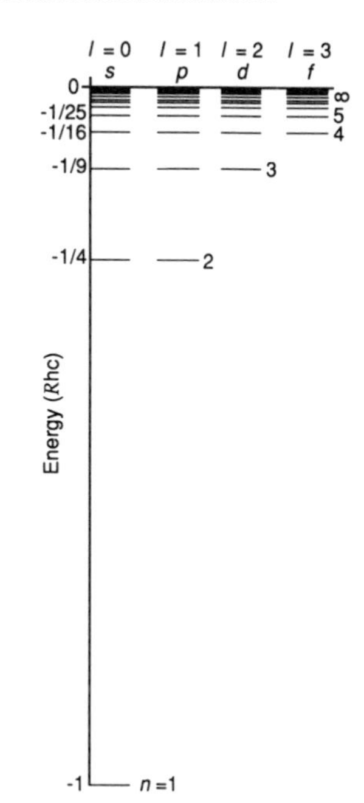

Fig. 2.9 The energy levels for the H-atom.

The magnetic quantum number m_l

The angular part of the wavefunction is characterized by a third quantum number m_l which is known as the magnetic quantum number. For every permitted value of l there are $2l + 1$ independent wavefunctions, with quantum numbers $m_l = -l, -l+1, \ldots, +l$. For the H-atom these wavefunctions all have the same energy and are known as *degenerate* states. Application of a magnetic or electric field to an atom leads to a lifting of the degeneracy of these states hence the name 'magnetic' (see Section 5.4).

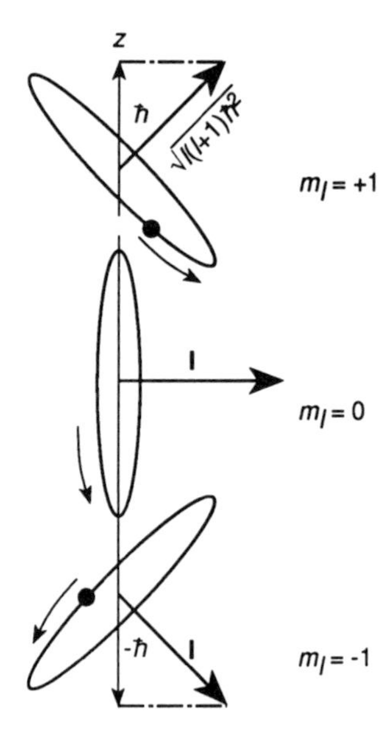

Fig. 2.10 The three space-quantized orientations for the angular momentum vector **l** of a state with $l = 1$. The length of **l** is $[l(l+1)\hbar^2]^{1/2}$.

The true meaning of this quantum number is that it gives directly the z-component of the orbital angular momentum vector, ℓ_z,

$$\ell_z = m_l \hbar. \tag{2.20}$$

The angular momentum vector for a p-state with $l = 1$ can have projections $\ell_z = +\hbar, 0, -\hbar$, as illustrated in Fig. 2.10. It should be re-emphasized that the states with these projections do not correspond directly to the common representation of p_x, p_y, or p_z orbitals (with lobes pointing along the x, y, z axes respectively). The p_x, p_y, p_z set are *real* functions whereas the $m_l = +1, 0, -1$ set are *complex* functions. In practice we are free to choose whichever representation of the p-orbitals is most appropriate; for atomic spectroscopy the $+1, 0, -1$ set is most useful, because there are selection rules governing the change of m_l, whereas in consideration of chemical bonding the p_x, p_y, p_z set is useful for discussion of the directional character of bonds.

The quantization of orientation is somewhat surprising given that the z-axis is fixed in the laboratory frame and the choice of its direction is arbitrary; we can change the orientation of the axes in our minds without affecting the atoms, yet this apparently changes the allowed projections of the angular momentum. The answer to this paradox lies in the fact that we cannot determine the state of an atom without performing an experiment such as the recording of a spectrum; as soon as this is done, the polarization vector of the light defines the axis system with respect to which the angular momentum is quantized and so the choice of axes is no longer arbitrary. In the absence of an experiment the true state of the system cannot be precisely specified, and the m_l quantum number of the electron is uncertain.

2.5 The solution of the Schrödinger equation

The Schrödinger equation for the hydrogen atom is normally solved using separation of variables by assuming a solution of the form

$$\Psi = R(r)Y(\theta, \phi). \tag{2.21}$$

Substituting this expression into eqn (2.4) gives

$$\frac{-\hbar^2}{2\mu}\left[\frac{1}{r}\frac{\partial^2}{\partial r^2}r\right]R(r)Y(\theta,\phi) - \frac{\hbar^2}{2\mu}\left[\frac{R(r)}{r^2}\Lambda^2 Y(\theta,\phi)\right] + V(r)R(r)Y(\theta,\phi) = ER(r)Y(\theta,\phi). \tag{2.22}$$

Of the operators on the left-hand side, only Λ^2 could change $Y(\theta, \phi)$ into some other function of the angular variables; therefore a valid solution can only be one for which this does not happen, i.e.,

$$\Lambda^2 Y(\theta, \phi) = CY(\theta, \phi), \tag{2.23}$$

where C is a constant. Making this substitution then leads to the angular parts dropping out of eqn (2.22) and we obtain a radial equation,

$$\frac{-\hbar^2}{2\mu}\left[\frac{1}{r}\frac{\partial^2}{\partial r^2}rR(r)+\frac{C}{r^2}R(r)\right]+V(r)R(r)=ER(r). \qquad (2.24)$$

The angular wave eqn (2.23) is of the same form as that for a particle on a sphere and it must be subject to the boundary conditions that

$$Y(\theta,\phi)=Y(\theta\pm 2n\pi,\phi\pm 2m\pi), \qquad (2.25)$$

where n, m are integers. The solutions of the angular equation are the spherical harmonics, listed in Table 1.2 (see Section 1.5). The constant C is restricted to the values $-l(l+1)$ where l is an integer

$$\Lambda^2 Y_{lm_l}(\theta,\phi)=-l(l+1)Y_{lm_l}(\theta,\phi). \qquad (2.26)$$

For every value of l, there are $2l+1$ independent solutions of the equation labelled by the quantum number m_l, which is the power to which $e^{i\phi}$ is raised in the spherical harmonic.

Making the substitution $C=-l(l+1)$ into eqn (2.24), and defining a new function $P(r)=r[R(r)]$, gives the radial equation of the form

$$\left[\frac{-\hbar^2}{2\mu}\frac{\partial^2}{\partial r^2}+V_{\text{eff}}\right]P(r)=EP(r) \qquad (2.27)$$

$$V_{\text{eff}}=\frac{-Ze^2}{4\pi\epsilon_0 r}+l(l+1)\frac{\hbar^2}{2\mu r^2}. \qquad (2.28)$$

V_{eff} is the *effective potential energy* in this one-dimensional equation, and it depends on l; it is the true potential energy plus a correction term for the centrifugal energy (see Fig. 2.11). The eqn (2.27) is in standard form, but it must be solved subject to the boundary condition that $R(r)\to 0$ as $r\to\infty$. The solutions are listed in Table 2.1 and the corresponding energy levels are given by

$$E=-\frac{\mu e^4}{32\pi^2\epsilon_0^2\hbar^2 n^2}=-\frac{\mathrm{R}hc}{n^2}, \qquad (2.29)$$

where n is restricted by the boundary conditions to the values 1,2,3. . . .

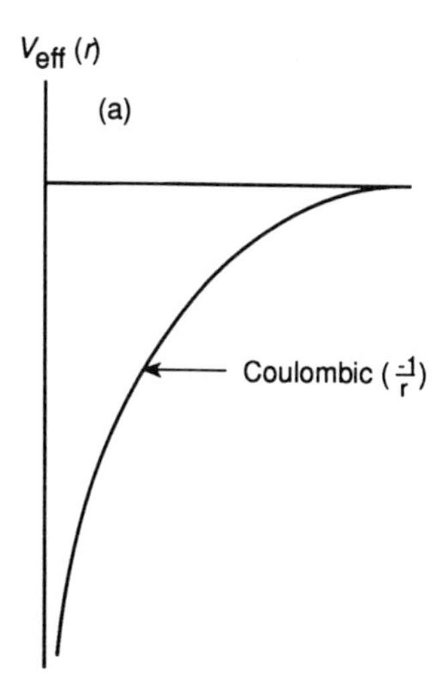

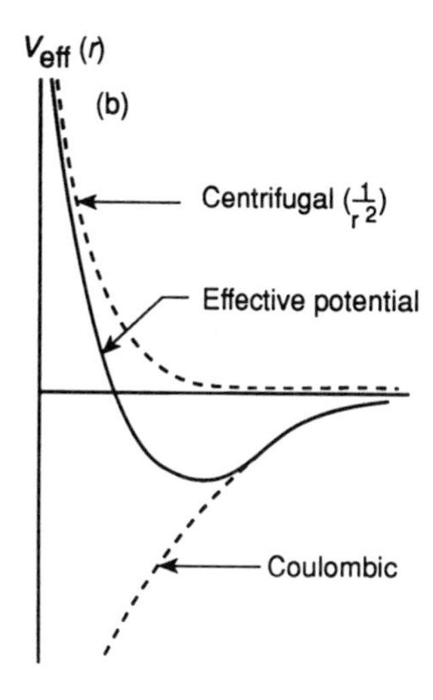

Fig. 2.11 The effective potential energy (eqn 2.28) for (a) $l=0$ and (b) $l\neq 0$. The effect of the centrifugal correction in (b) is to prevent the electron existing at $r=0$ where the effective potential is infinite.

2.6 Population factors

The appearance of any atomic spectrum is determined not only by the underlying energy levels but also by *population factors* and *selection rules*. An absorption or emission of radiation involving a jump from initial energy level i to final energy level f can only occur if there are a significant number of atoms in the sample existing in state i. In some cases, for example in a room temperature sample of helium gas, all the atoms would exist in the lowest energy level. Under these circumstances, emission of radiation does not occur, and only an absorption from the lowest level to a higher level is possible. For a sample of gas under conditions of thermal equilibrium, the fraction of atoms in level i, n_i/N, with energy E_i is given by the Boltzmann expression,

$$\frac{n_i}{N} = \frac{g_i \exp(-E_i/kT)}{\sum_i g_i \exp(-E_i/kT)}, \tag{2.30}$$

where g_i is the degeneracy of level i, N is the total number of atoms and k is the Boltzmann constant. At equilibrium for a temperature of 300 K the fraction of atoms in the second energy state of helium, whose energy is 3.2×10^{-18} J above the ground state is calculated to be $\sim 10^{-348}$, justifying the statement above. The spectrum of the hydrogen atom is normally recorded in a discharge tube where the effective temperature is many thousands of Kelvin. The atoms exist in a wide range of energy levels and emission can be observed from almost any level.

The amount of light emitted or absorbed by an atom can generally be related quantitatively to the population of the initial state of the transition—the greater the population, the greater the intensity of light emitted or absorbed. For absorption this is expressed in the Beer–Lambert law,

$$I = I_0 \exp(-\kappa c l), \tag{2.31}$$

where I is the intensity transmitted through a sample, I_0 is the incident intensity, l is the length of the sample and c is the concentration of the absorbing species in the initial state of interest. κ is known as the molar absorption coefficient and a measure of the probability of the transition occurring (see Section 2.8 $\kappa cl \ll 1$ then $I \sim I_0(1 - \kappa cl)$ or

$$(I_0 - I) \sim \kappa cl\, I_0. \tag{2.32}$$

The intensity of light absorbed, $I_0 - I$, is proportional to the concentration of absorbing species.

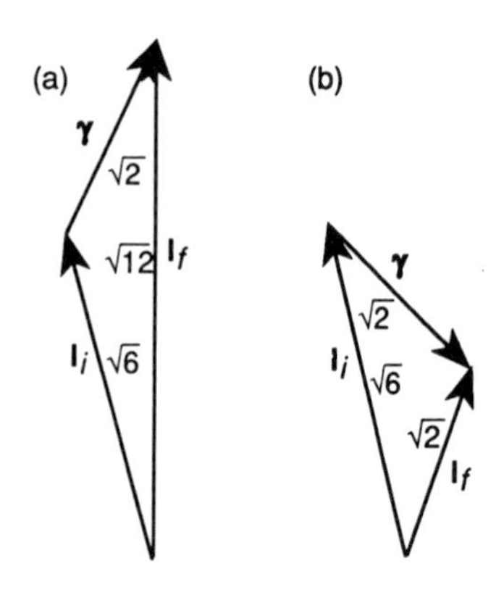

Fig. 2.12 Conservation of orbital angular momentum on absorption of a photon. The sum of the photon angular momentum vector γ and initial orbital angular momentum $\mathbf{l}_i$ must give a resultant $\mathbf{l}_f$ of magnitude $[l'(l'+1)]^{\frac{1}{2}}\hbar$ where l' is integral. The maximum (a) and minimum (b) resultants for $l_i = 2$ are shown, with lengths given in units of $\hbar$.

2.7 Selection rules

Not all energy jumps are possible in an atom even if the population factors are sufficient; *selection rules* are a statement of how the quantum numbers are permitted to change. For the hydrogen atom these may be written:

$$\begin{aligned} &\Delta n \text{ no restriction,} \\ &\Delta l = \pm 1 \text{ only,} \\ &\Delta m_l = 0, \pm 1. \end{aligned} \tag{2.33}$$

The $\Delta l = \pm 1$ selection rule is the most significant here and arises from two factors; conservation of angular momentum, and symmetry conservation. Although a detailed explanation is beyond the scope of this text, we can understand why Δl cannot be greater than one with the knowledge that a photon carries with it one unit of angular momentum, and that angular momentum must be conserved in the transition. Thus, if we add the angular momentum of the photon γ to the initial angular momentum of the atom $\mathbf{l}_i$, the final angular momentum of the atom $\mathbf{l}_f$ cannot increase by more than one unit.

More correctly we should consider the angular momentum as a *vector* quantity; it is the vector sum that must be conserved

$$\mathbf{l}_i + \gamma = \mathbf{l}_f. \tag{2.34}$$

The vector $\mathbf{l}_i$ has magnitude $[l(l+1)]^{\frac{1}{2}}\hbar$ and it must add with the photon angular momentum vector γ whose magnitude is $\sqrt{2}\hbar$ (equivalent to an angular momentum quantum number of 1) to give a final angular momentum of magnitude $[l'(l'+1)]^{\frac{1}{2}}\hbar$ where l' must be integral. It is easily shown that the maximum quantum number is $l' = l + 1$ and the minimum is $l' = l - 1$. The vector addition is illustrated in Fig. 2.12.

Symmetry constraints

Angular momentum conservation does *not* rule out the $\Delta l = 0$ process, but this is actually forbidden by a *symmetry* conservation rule. The angular wavefunctions for some of the s, p and d orbitals are shown in Fig. 2.13. It can be seen clearly that the s-wavefunction is spherically symmetric and is therefore an *even* function with respect to inversion of the *x,y,z* coordinates; i.e., the value of the wavefunction at (X, Y, Z) is the same as at $(-X, -Y, -Z)$. The p-wavefunction on the other hand is an *odd* function (antisymmetric) with respect to inversion of the coordinates, that is, if we take the value of the wavefunction at any point in space, and then compare it with the value at a position on the other side of the origin diametrically opposite, the wavefunction has the same magnitude but changes sign. The d-wavefunction is again symmetric (even) with respect to inversion, because the value of the wavefunction at the diametrically opposite point always has the same sign.

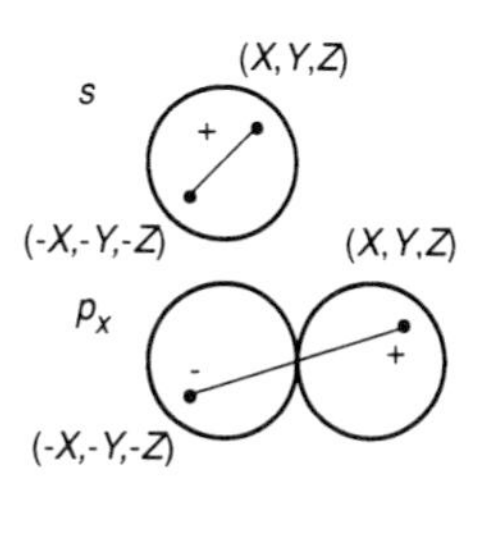

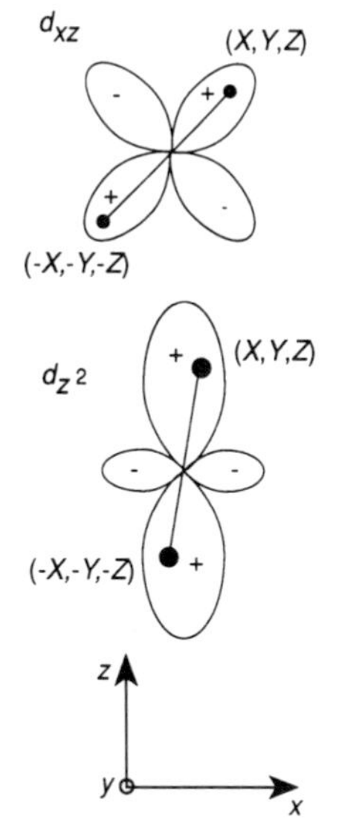

Fig. 2.13 The s and d orbitals are symmetric with respect to inversion of coordinates through the origin, whereas the p orbitals are antisymmetric.

The electromagnetic field is represented by an *odd* function in eqn (1.3) (antisymmetric with respect to inversion). As a consequence, it imposes a change in the inversion symmetry of the atomic wavefunction when a photon is absorbed or emitted. (This point is put on a quantitative footing in the next section.) Therefore, a $\Delta l = \pm 1$ transition such as $s \to p$ or $p \to d$ is allowable, but $s \to d$ or $s \to s$ do not change symmetric to antisymmetric and so these are forbidden. To summarize, the change of l must be odd-numbered.

2.8 Quantum mechanical formulation of selection rules

Time-dependent quantum mechanics can be used to calculate the rate at which an atom in level i with wavefunction Ψ_i undergoes a transition to a level f with wavefunction Ψ_f, when subject to a sinusoidally oscillating electric field (i.e., a plane electromagnetic wave). The result obtained is

$$W_{if} = B_{if} I(\nu) c \qquad B_{if} = \frac{1}{6\epsilon_0 \hbar^2} \left| \int \Psi_i^* \hat{\mu} \Psi_f \mathrm{d}\tau \right|^2. \tag{2.35}$$

B_{if} is known as the Einstein coefficient of absorption, $\int \Psi_i^* \hat{\mu} \Psi_f \mathrm{d}\tau$ is the *transition moment*, and $\hat{\mu}$ is the dipole moment operator $-e\hat{\mathbf{r}}$. W_{if} is the rate of change of the probability that the atom is in state f. The selection rules referred to in Section 2.7 are a statement of under what conditions the transition moment integral is zero or nonzero. A *forbidden transition* is one for which the integral is zero. Thus, the symmetry constraints on the transition can be understood more easily by realizing that the integral over all space of an antisymmetric (odd) function must be zero. Therefore the product $\Psi_i^* \hat{\mu} \Psi_f$ must be an *even* function overall for the transition to be possible. The dipole moment is an odd function and using the rules for multiplication of functions

$$\begin{aligned} odd \times odd &= even \\ even \times even &= even \\ odd \times even &= odd, \end{aligned} \tag{2.36}$$

it is clear that allowed transitions will have either Ψ_i or Ψ_f as an odd function, but not both.

The angular momentum constraint, restricting the maximum change of l to 1 cannot be demonstrated so easily. In principle, explicit substitution of the angular parts of the wavefunction into eqn (2.35) would show that the integral is always zero when $\Delta l \geq 2$, hence the selection rule $\Delta l = \pm 1$.

Selection rules tell us whether a transition can happen or not, but do not give information on how probable the transition is; this information is contained in the transition rate expression (2.35). It is useful to be able to draw a connection between the experimentally determined intensity of light absorbed, $I - I_0$, with the rate expression (2.35). The molar absorption coefficient $\kappa\ (\nu)$ is the experimental measure of transition probability, and is defined as the proportionality constant in the equation giving the change in the intensity of light, dI passing through a sample of concentration c and length dx

$$\mathrm{d}I = -I\kappa c \mathrm{d}x. \tag{2.37}$$

Integrating this equation between $x = 0$ and l gives the Beer–Lambert law (in a slightly different form from eqn 2.31),

$$\ln \frac{I}{I_0} = -\alpha \kappa c l, \tag{2.38}$$

where I_0 and I are the intensities at $x = 0$ and $x = l$ respectively. As discussed elsewhere, absorption lines always have a finite *linewidth*, and the transition probability is spread out over that region. The integrated absorption probability $\mathcal{A}$ is given by

$$\mathcal{A} = \int \kappa(\nu) \mathrm{d}\nu, \tag{2.39}$$

where the integral is over the range of frequencies of the line. $\mathcal{A}$ is related to the theoretical Einstein coefficient of absorption B_{if} by

$$\mathcal{A} = \frac{h\nu L B_{if}}{c}. \tag{2.40}$$

L is Avogadro's number, and ν is the transition frequency at the centre of the line.

2.9 The spectrum of the hydrogen atom

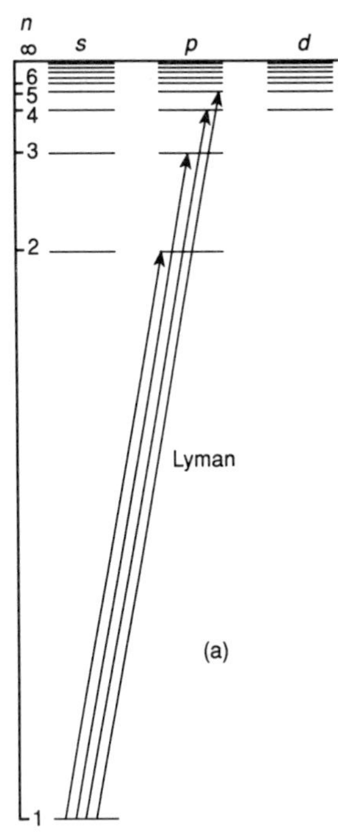

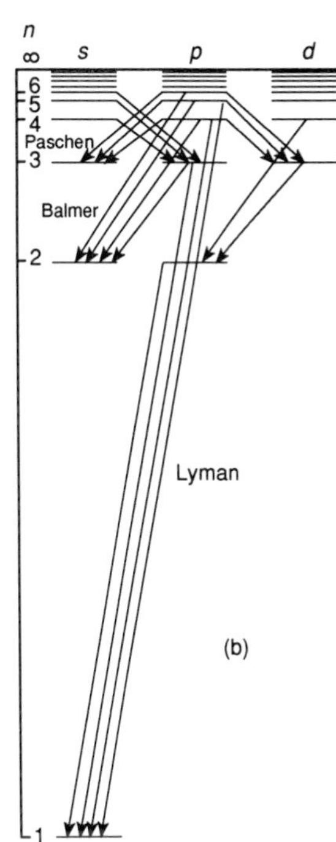

Fig. 2.14 Grotrian diagram showing some of the permitted transitions in (a) absorption and (b) emission.

In Fig. 2.14a the allowed transitions in *absorption* involving the $1s$ ground state of the H-atom are shown in a *Grotrian diagram*. Only transitions to the np levels are allowed by the $\Delta l = \pm 1$ selection rule. The $n = 1$ to $n = 2$ energy gap is very much larger than any other energy gap; consequently, the transitions to and from the ground state involve much bigger energy jumps than any other transitions. The light emitted or absorbed appears in the far ultraviolet region, and this part of the H-atom spectrum is known as the Lyman series. Although there are many different energy levels populated in the hydrogen discharge, the overwhelming majority of atoms are in the ground state; even at a temperature of 10 000 K the Boltzmann population of the $n = 2$ level is smaller than the ground state by a factor of 6×10^{-6}; therefore only the Lyman series is normally observed in absorption. However, absorption lines originating from $n = 2$ levels have been observed in the light emitted by stars where the temperature is very high.

The transitions with the $n = 2$ levels as the lower state are known as the Balmer series and occur in the visible region of the *emission* spectrum, as illustrated in Fig. 2.14b. According to the selection rules these could be $np \to 2s$, $nd \to 2p$ or $ns \to 2p$. As the levels of given n but different l are degenerate in the present level of approximation, the three types of transition are superimposed (see Fig. 1.2a).

The transition energies are simply given by the differences between levels of different n

$$\Delta E = E_i - E_f = h\nu = \frac{hc}{\lambda} = -\mathcal{R}hc\left(\frac{1}{n_i^2} - \frac{1}{n_f^2}\right) \tag{2.41}$$

with the final-state quantum number $n_f = 2$ for the Balmer series.

Ionization energy

The ionization energy of the hydrogen atom is, by definition, the energy required to remove an electron from the atom in its ground state to an infinite distance from the nucleus. This is simply equal to the negative of the $1s$ electron energy

$$\text{Ionization energy} = \frac{+\mathcal{R}hc}{n^2} = \mathcal{R}hc \quad \text{for} \quad n = 1. \tag{2.42}$$

It corresponds to the transition energy required to excite an electron to a state of infinite principal quantum number. Looking at Fig. 2.3 it can be seen that the observed transition frequencies in the Lyman series get closer and closer together as n_i increases. In principle, the ionization energy can be determined by the limit of the series where the lines become a continuum. Above the ionization limit the electron is unbound and does not have quantized energy, hence there is a continuous absorption probability. In practice it may not be easy to determine the convergence accurately and a more precise method is to plot the transition energies versus $1/n_i^2$. n_f is constant for a particular series

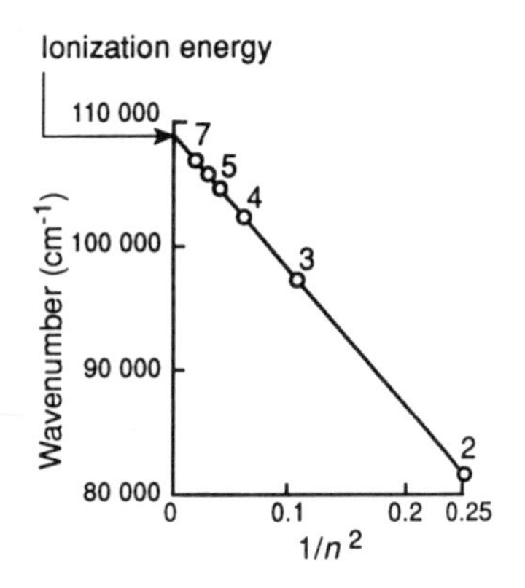

Fig. 2.15 Graphical procedure for determination of the ionization potential of the hydrogen atom from the wavenumbers of lines in the Lyman series. The wavenumbers are plotted versus $1/n^2$ where n is the upper state quantum number. The intercept gives the ionization energy.

and so a straight line is obtained of gradient $-\mathcal{R}hc$. The intercept is measured at $1/n_i^2 = 0$ and this is equal to $\mathcal{R}hc/n_f^2$ from which the ionization energy is extracted. The procedure is illustrated in Fig. 2.15.

Hydrogen-like ions

The energy levels determined from the Schrödinger equation for the isoelectronic ions, He^+, Li^{2+} ... are given by the same expression as for the hydrogen atom but with an additional factor of Z^2

$$E = \frac{-Z^2\mathcal{R}hc}{n^2}. \tag{2.43}$$

This factor originates from the inclusion of the atomic number Z in the potential energy part of the Schrödinger equation (eqn 2.8). Thus the energy level spacings in helium (1+) are four times those in the H atom, those for lithium(2+) are nine times as large and so on, and so the frequencies of the observed transitions differ by the same factor. In fact this relationship is not exact because the Rydberg constant differs slightly for the various ions through its dependence on the electron–nucleus reduced mass (eqns 2.2 and 2.3). The values of $\mathcal{R}$ for various ions are shown in Table 2.2. An interesting consequence of the Z^2 dependence of the energies is that the transition frequencies for the $1 \to 2$, $1 \to 3$, $1 \to 4$ transitions in the H-atom are almost degenerate with the transitions $2 \to 4$, $2 \to 6$, $2 \to 8$ in He^+.

2.10 Spin–orbit coupling

Although the above description is adequate for describing the gross features of the hydrogen atom spectrum, it fails to account for the very small splittings of the lines that are observed using high-resolution spectrometers. For example, Fig. 2.16 shows that the $n = 3$ to $n = 2$ emission line actually appears as seven closely-spaced components, separated into two groups. The most important effect causing deviations from the predictions of the Schrödinger equation is *spin–orbit coupling*. To understand this effect, a discussion of electron spin is necessary.

Table 2.2
Rydberg constants for hydrogenic atoms

Atom	$\mathcal{R}$ (cm^{-1})
H	109 677.8
He^+	109 722.3
Li^{2+}	109 728.7
Be^{3+}	109 730.6

Spin quantum numbers for an electron

Electrons and many other types of elementary particle can be considered to possess an intrinsic angular momentum; the charged particle spins about its own axis and the property is known as *spin*. The magnitude of the spin angular momentum of an electron, represented by the vector $\mathbf{s}$, is always $\sqrt{3}\hbar/2$ and therefore by analogy with our description of the orbital angular momentum we define a *spin quantum number* $s = \frac{1}{2}$ such that $|\,\mathbf{s}\,|$ is given by

$$|\,\mathbf{s}\,| = \sqrt{s(s+1)}\hbar = \frac{\sqrt{3}}{2}\hbar. \tag{2.44}$$

s is a rather unusual quantum number in that it can only take one value. Just as the orbital angular momentum has quantized z-components, so the projection of the spin angular momentum s_z is restricted to having values

$$s_z = \pm\frac{\hbar}{2} \tag{2.45}$$

and the m_s quantum number is defined to have values $\pm\frac{1}{2}$ such that $s_z = m_s\hbar$.

The property of electron spin was first proposed by Uhlenbeck and Goudsmid in 1925 and confirmed by the famous Stern–Gerlach experiment. The experiment consisted of passing a beam of silver atoms through an inhomogeneous magnetic field as shown in Fig. 2.17, and it was discovered that the silver atoms traversed two distinctive paths through the field. This behaviour was ascribed to the outermost electron in the atom having a spin quantum number $m_s = \pm\frac{1}{2}$ but with no orbital angular momentum $l = 0$. The spin of the charged particle gave it a magnetic moment and hence an interaction with the external magnetic field that depended on its orientation (see Section 5.4). The two spin quantum states of the atom therefore experienced different forces in the inhomogeneous field and the atoms ended up spatially separated. It was the observation that there were only *two* groups of atoms that told the experimenters they were dealing with an angular momentum with associated quantum number $\frac{1}{2}$. In principle, this experiment could have been performed with hydrogen atoms but at the time the technology for producing H-atom beams was not available.

Theoretically, the Schrödinger equation does *not* predict the existence of electron spin. However, Dirac showed in 1928 that if the ideas of wave mechanics were extended to be fully consistent with Einstein's theory of special relativity, then a new equation, later known as the Dirac equation, must be written down to replace the Schrödinger equation, containing a fourth coordinate, time. A fourth quantum number then becomes necessary to label the four dimensional wavefunctions and this is identified as the magnetic spin quantum number m_s. In fact, there is no real evidence that the electron does actually spin in the classical sense, although it certainly does have a magnetic moment;

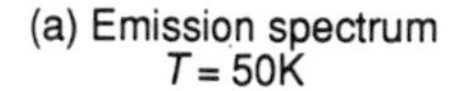

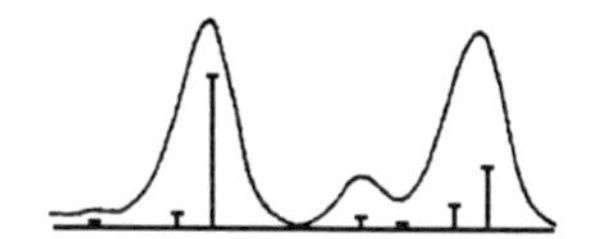

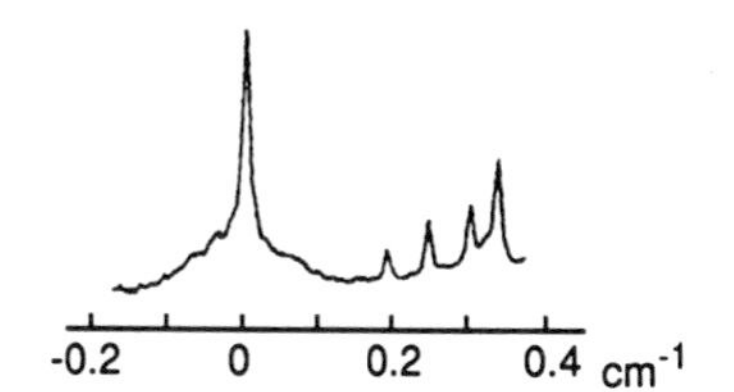

Fig. 2.16 (a) Emission spectrum showing some substructure for the $n = 3 \rightarrow 2$ line of hydrogen at a temperature of 50 K; the low temperature helps to reduce the Doppler broadening of lines. The positions of the seven underlying fine-structure components are indicated with bars. (b) High resolution laser 'saturation spectrum' showing full resolution of the fine-structure components. The strong line $\left(\mathrm{P}_{\frac{3}{2}} - \mathrm{D}_{\frac{5}{2}}\right)$ is at 15 233.0702 cm^{-1}, and the horizontal scale is relative to this wavenumber.

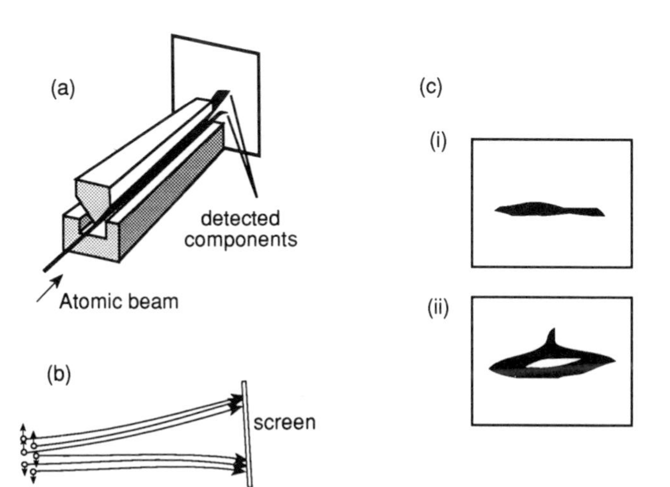

Fig. 2.17 (a) Schematic diagram of the Stern–Gerlach experiment. (b) Trajectories of silver atoms are divided into two groups, corresponding to spin states $m_s = +\frac{1}{2}$ and $m_s = -\frac{1}{2}$. (c) The pattern observed on the screen
(i) with no magnetic field
(ii) with magnetic field.

the Dirac equation does not predict spin as such. Nevertheless, we will continue to use the idea of an intrinsic angular momentum, because of the useful mathematical and pictorial developments that arise from it.

Spin–orbit coupling

As stated above the intrinsic spin of an electron leads to this particle acting like a tiny bar magnet, with magnetic moment $\mathbf{m}_s$ given by:

$$\mathbf{m}_s = g_e \gamma_e \mathbf{s} \qquad \gamma_e = -\frac{e}{2m_e}. \tag{2.46}$$

g_e is the electronic g-factor (a constant) which has been accurately measured to the value 2.002 319 314, while γ_e is the magnetogyric ratio. The orbital angular momentum possessed by an electron with $l > 0$ leads to a second source of magnetism. The circulating charged particle is an electric current and produces at the centre of the orbit a magnetic moment $\mathbf{m}$ given by:

$$\mathbf{m} = \gamma_e \mathbf{l}. \tag{2.47}$$

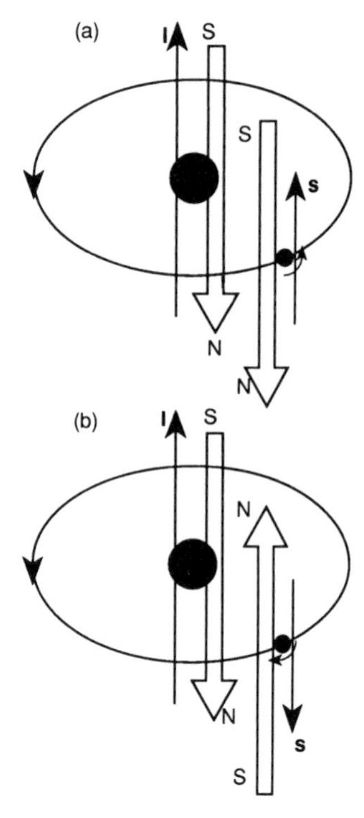

Fig. 2.18 (a) The orbital angular momentum vector **l** and spin angular momentum vector **s** are parallel; hence the two magnetic moments are also parallel (in the opposite direction) and repel resulting in a high energy state. (b) The magnetic moments are antiparallel and there is an attractive low-energy interaction. The north (N) and south (S) poles indicate the analogy with two simple bar magnets.

Just as two bar magnets repel or attract one another according to their relative orientation, so the orbital and spin magnetic moments can be lined up in an attractive or repulsive orientation, that is energetically favourable or unfavourable respectively, as shown in Fig. 2.18.

Classically the energy of interaction ϵ between two magnetic dipoles $\mathbf{m}_1$ and $\mathbf{m}_2$ separated by the vector $\mathbf{r}$ (see Fig. 2.19) has the form:

$$\epsilon = \frac{\mu_0}{4\pi r^3}[\mathbf{m}_1.\mathbf{m}_2 - 3\mathbf{m}_1.\hat{\mathbf{r}}\ \hat{\mathbf{r}}.\mathbf{m}_2], \tag{2.48}$$

where $\hat{\mathbf{r}}$ is the unit vector along the direction of $\mathbf{r}$. The energy depends strongly on the separation and relative orientation of the dipoles. We have seen, however, that in quantum mechanics, energies are not continuous but quantized. It should be no surprise therefore to learn that the spin and orbital angular momenta cannot take up any relative orientation, but only certain permitted orientations, and hence only certain permitted energies of interaction.

It is convenient to define a *total angular momentum* vector **j** as the vector sum of **l** and **s**

$$\mathbf{j} = \mathbf{l} + \mathbf{s}. \tag{2.49}$$

Quantum mechanically, the vector **j** behaves in a very similar way to **l**; in particular the operators representing j^2 and j_z have the same commutation properties (see Section 1.5) as those representing ℓ^2 and ℓ_z. The magnitude of **j** is restricted to values $[j(j+1)]^{\frac{1}{2}}\hbar$ where j is the total angular momentum quantum number, analogous to l. j is restricted to have values which are half-integral for the hydrogen atom as follows:

$$j = l + s, l + s - 1, \ldots, \mid l - s \mid. \tag{2.50}$$

As $s = \frac{1}{2}$ there are only two possible values, namely $j = l \pm \frac{1}{2}$, which represent two different orientations of the spin and orbital angular momenta. These two orientations (almost as shown in Figs 2.18 (a) and (b) respectively) represent the two magnetic moments being aligned almost parallel and antiparallel and therefore represent two different quantized energy states. Thus, an electron in a $2p$ orbital of the hydrogen atom has two possible energy states with $j = \frac{1}{2}$ and $j = \frac{3}{2}$, the latter being of higher energy. A $3d$ electron can have $j = \frac{3}{2}$ and $j = \frac{5}{2}$. An ns electron has no orbital angular momentum and there is no spin–orbit coupling. The total angular momentum is simply equal to the spin angular momentum, and therefore $j = \frac{1}{2}$ only.

These features are illustrated in Fig. 2.20, which shows the $n = 2$ and 3 energy levels of the hydrogen atom with inclusion of spin–orbit coupling. (The 'term symbol' notation used for the energy levels is explained at the end of this section.) It should be noted that the energy level splittings (considered in detail below) are different for different n and l.

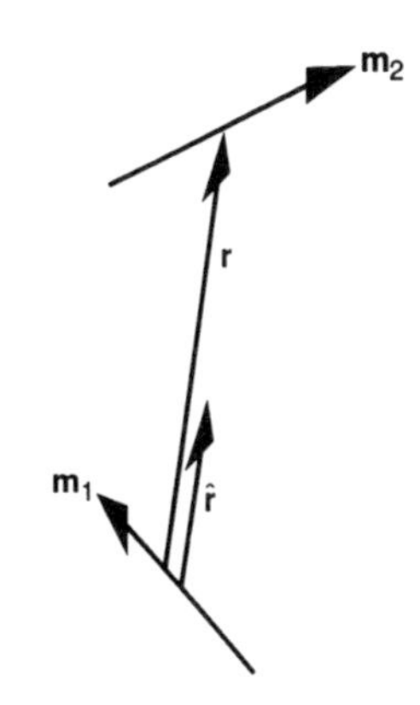

Fig. 2.19 Two magnetic dipole moment vectors $\mathbf{m}_1$ and $\mathbf{m}_2$, separated by the vector $\mathbf{r}$. $\hat{\mathbf{r}}$ is a unit vector along the direction of $\mathbf{r}$.

Additional selection rules

When spin–orbit coupling is taken into account, an additional selection rule that applies in the absorption and emission spectrum of the hydrogen atom is

$$\Delta j = 0, \pm 1. \tag{2.51}$$

Consider first the Lyman series, $np \to 1s$. Both transitions $np(j = \frac{3}{2}) \to ns(j = \frac{1}{2})$ and $np(j = \frac{1}{2}) \to ns(j = \frac{1}{2})$ are permitted by the above selection rule and the observed lines are split into two components known as a *doublet*. The energy splitting between the two lines of the doublet is equal to the energy splitting of the np level. For the Balmer series, the allowed transitions, illustrated in Fig. 2.20 are as follows:

$$np\left(j = \frac{3}{2}\right) \to 2s\left(j = \frac{1}{2}\right) \quad np\left(j = \frac{1}{2}\right) \to 2s\left(j = \frac{1}{2}\right)$$

$$ns\left(j = \frac{1}{2}\right) \to 2p\left(j = \frac{3}{2}\right) \quad ns\left(j = \frac{1}{2}\right) \to 2p\left(j = \frac{1}{2}\right)$$

$$nd\left(j = \frac{5}{2}\right) \to 2p\left(j = \frac{3}{2}\right) \quad nd\left(j = \frac{3}{2}\right) \to 2p\left(j = \frac{3}{2}\right)$$

$$nd\left(j = \frac{3}{2}\right) \to 2p\left(j = \frac{1}{2}\right). \tag{2.52}$$

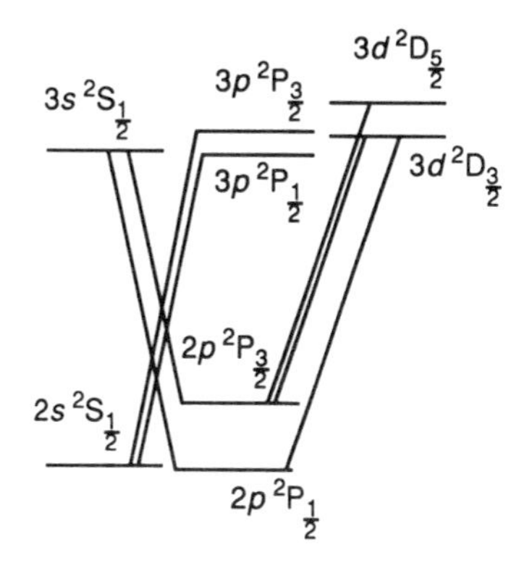

Fig. 2.20 The fine-structure splitting for the $n = 2$ and 3 levels of hydrogen (not to scale). The seven transitions shown in Fig. 2.16 are indicated.

One transition is forbidden, $nd(j = \frac{5}{2}) \to 2p(j = \frac{1}{2})$ because this would involve $\Delta j = -2$. It is shown in Fig. 2.16 that these transitions lead to the observation of seven lines instead of one in the spectrum. The splittings of all levels except the $2p$ are very small and consequently the lines are grouped into two sets as shown in Fig. 2.16; one set of four lines involves the lower states $2p(j = \frac{1}{2})$ and $2s(j = \frac{1}{2})$, while the group of three lines are transitions to the $2p(j = \frac{3}{2})$ level.

Term symbols

It is convenient to use a notation known as the *term symbol* to label the states as shown in the right-hand column below

$$
\begin{aligned}
&3s\left(j=\frac{1}{2}\right) &\qquad& 3s\left({}^2\mathrm{S}_{\frac{1}{2}}\right)\\
&2p\left(j=\frac{3}{2}\right) && 2p\left({}^2\mathrm{P}_{\frac{3}{2}}\right)\\
&3d\left(j=\frac{5}{2}\right) && 3d\left({}^2\mathrm{D}_{\frac{5}{2}}\right).
\end{aligned}
\tag{2.53}
$$

The superscript 2 is the so-called *spin multiplicity*. For a one-electron atom the electron has two possible spin states, $m_s = \frac{1}{2}$ and $m_s = -\frac{1}{2}$, and so the spin multiplicity is 2. The orbital angular momentum is rewritten as a capital letter, S, P, D and j is written as a subscript. The usefulness of this notation becomes more apparent when we deal with many-electron atoms.

2.11 Quantitative description of spin–orbit coupling

A rigorous derivation of the spin–orbit coupling energy can only be obtained by reverting to the relativistic formulation of quantum mechanics. The Schrödinger equation does not predict spin–orbit coupling because it ignores electron spin; however, this interaction is fully predicted by the relativistic Dirac equation. The mathematics involved is well beyond the scope of this book; for the present purposes we can gain insight by adopting a classical model and then jump to the quantum mechanical results. A convenient starting point is to turn the atom 'inside out' and consider a positively charged nucleus orbiting around a stationary (but spinning) electron, with angular velocity

$$\omega = \frac{\mathbf{l}}{m_e r^2}, \tag{2.54}$$

(r is the mean orbital radius and $\mathbf{l}$ is the angular momentum vector). This is a legitimate frame transformation that simplifies the analysis: the orbital magnetic field is then a vector that originates from the position of the electron, at the centre of the orbit, and it is easier in this way to calculate the interaction energy with the spin magnetic moment which is a vector with the same origin. The electric current i is equal to $\omega Ze/2\pi$ and this gives a magnetic flux density $\mathbf{B}$ at the centre of the loop of magnitude

$$|\mathbf{B}| = \frac{\mu_0 i}{2r} = 4\pi \times 10^{-7}\left[\frac{|\mathbf{l}|Ze}{4\pi m_e r^3}\right], \tag{2.55}$$

where μ_0 is the vacuum permeability. The classical energy of interaction between the spin magnetic moment (eqn 2.46) and the magnetic field $\mathbf{B}$ is then given by

$$E = -\mathbf{m}.\mathbf{B}, \tag{2.56}$$

leading to

$$E = \frac{-\mu_0 g_e \gamma_e Ze}{4\pi m_e r^3}\mathbf{l}.\mathbf{s}. \tag{2.57}$$

Finally using the mean value of $\frac{1}{r^3}$ for an electron in an orbit with quantum numbers n, l

$$\langle 1/r^3\rangle_{nl} = \frac{(Z/a_0)^3}{n^3 l(l+\frac{1}{2})(l+1)} \qquad a_0 = \frac{4\pi\epsilon_0\hbar^2}{m_e e^4}, \tag{2.58}$$

we find that the spin–orbit energy is

$$E_{\text{so}} = \xi\mathbf{l}.\mathbf{s}, \tag{2.59}$$

where ξ is a constant depending on n, l and Z^4. The classical model does not quite give the correct value for the spin–orbit coupling constant although it does predict the correct n, l and Z dependence. The full quantum mechanical theory from relativistic quantum mechanics predicts the spin–orbit energy is given by

$$\hbar^2 E_{\text{so}} = hc\zeta_{nl}\langle\mathbf{l}.\mathbf{s}\rangle \qquad \zeta_{nl} = \frac{\alpha^2 \mathcal{R} Z^4}{n^3 l(l+\frac{1}{2})(l+1)}, \tag{2.60}$$

with $\mathcal{R}$ the Rydberg constant, and α the so-called fine structure constant given by

$$\alpha = \frac{e^2}{4\pi\epsilon_0\hbar c}. \tag{2.61}$$

The expectation value $\langle\mathbf{l}.\mathbf{s}\rangle$ is evaluated using $\mathbf{j} = \mathbf{l} + \mathbf{s}$, hence, $\mathbf{j}^2 = \mathbf{l}^2 + \mathbf{s}^2 + 2\mathbf{l}.\mathbf{s}$ and therefore,

$$\begin{aligned} E_{\text{so}} &= \frac{1}{2}\langle\zeta(\mathbf{j}^2 - \mathbf{l}^2 - \mathbf{s}^2)\rangle\frac{hc}{\hbar^2} \\ &= \frac{1}{2}\zeta[\langle\mathbf{j}^2\rangle - \langle\mathbf{l}^2\rangle - \langle\mathbf{s}^2\rangle]\frac{hc}{\hbar^2} \\ &= \frac{1}{2}\zeta[j(j+1) - l(l+1) - s(s+1)]hc. \end{aligned} \tag{2.62}$$

To a good approximation (first-order perturbation theory—see the Background reading), the spin–orbit energy (2.62) can be simply added to the energy derived from the Rydberg formula (2.29) to obtain the energy levels.

Table 2.3 lists the observed magnitudes of the spin–orbit splitting between levels $j = l \pm \frac{1}{2}$ for an electron in various orbitals of the hydrogen atom, and also the $2p$ splittings for H, He^+, Li^{2+}, Be^{3+}, and B^{4+}. These data confirm the trends predicted by the theory above, i.e., the spin–orbit splitting decreases with n and l and is proportional to Z^4.

Table 2.3
Spin–orbit splittings for hydrogenic orbitals

Atom	Orbital	Spin–orbit splitting (cm^{-1})
H	$2p$	0.365
H	$3p$	0.108
H	$3d$	0.036
H	$4p$	0.046
H	$4d$	0.015
H	$4f$	0.008
He^+	$2p$	5.843
Li^{2+}	$2p$	29.58
Be^{3+}	$2p$	93.5
B^{4+}	$2p$	228.3

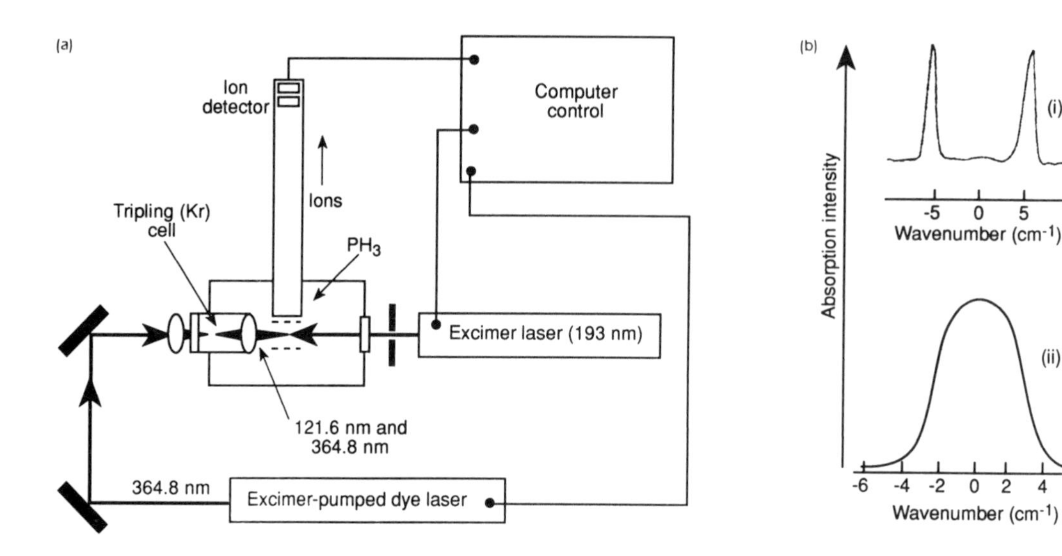

Fig. 2.21
(a) Experimental arrangement for photodissociation studies: the excimer laser produces intense monochromatic radiation at 193 nm, photodissociating PH_3 in the cell. The excimer-pumped dye laser plus tripling cell produces light at two wavelengths, 364.8 nm and 121.6 nm. The latter wavelength excites the $n = 2 \leftarrow 1$ transition of hydrogen, while the former ionizes the excited $n = 2$ atoms to produce H^+ which is then detected in a mass spectrometer. This procedure is a highly sensitive method for measuring the absorption by the H-atom. (b) The line profile of the $n = 2 \leftarrow n = 1$ absorption line for the H-atoms produced by photodissociation of (i) HBr (ii) PH_3 (wavenumbers are given with respect to the line centre).

2.12 An application in chemical reaction dynamics

An important area of chemical physics known as *chemical reaction dynamics* attempts to explore the details of chemical reactions and dissociation processes on a microscopic molecular level. An interesting application of the atomic spectrum of hydrogen in this field is shown in Fig. 2.21. An intense source of light from an excimer laser interacts with a sample of phosphine gas (PH_3), and puts sufficient energy into the molecules to cause bond dissociation (photodissociation)

$$PH_3 \rightarrow PH_2 + H.$$

An important aid to understanding the dynamics of this process would be to determine how much of the excess energy (i.e., the photon energy minus the dissociation energy) put into a molecule is carried away as kinetic energy of the hydrogen atom and hence how much remains as internal energy of the PH_2 fragment. Also of interest is to find out whether the hydrogen atoms are thrown out in a direction parallel to the laser polarization, or perpendicular, or with no preference. These features of the dissociation process can be determined by recording the absorption spectrum of the ejected hydrogen atoms *in situ* immediately following photodissociation. The absorption spectrum is obtained using another laser as the light source which can be accurately directed into the small volume where photodissociation has occurred.

The required information about energy distribution discussed above is obtained by careful study of the absorption lineshapes. Real spectroscopic absorption features are not infinitesimally narrow but have a finite linewidth caused by a number of factors. One of these is the Doppler effect (see Section 1.2), in which atoms moving towards the light source observe a higher effective frequency of the light and therefore absorb at lower actual frequencies compared to those moving away from the light source. This leads in general to a distribution of actual transition frequencies depending on the distribution of molecular velocities. Figure 2.21b shows the profile of the Lyman α line ($n = 2 \leftarrow n = 1$)

recorded for H-atoms produced by photodissociation of PH_3 and compares it with the profile of the same spectroscopic transition for the dissociation of HBr under similar conditions. The two peaks in the latter profile are due to energetic atoms moving forwards or backwards with respect to the direction of the laser beam measuring absorption. The separation between the peaks can be used to determine the mean kinetic energy of the ejected H-atoms and the overall profile reflects the angular distribution of the H-atom fragment recoil relative to the laser direction. The H-atoms from PH_3 photodissociation show a narrower overall line profile peaked at zero offset relative to the true transition frequency. The Doppler profile can be analysed to infer that much of the excess energy of the photodissociation is partitioned into vibration of the PH_2 molecule, rather than into kinetic energy of the H-atom.

2.13 Problems

1. The table below lists the wavenumbers of some of the transitions in (i) the emission spectrum of atomic hydrogen and (ii) the absorption spectrum of an unknown hydrogenic ion (e.g., He^+, Li^{2+} ...).

(i) Hydrogen (cm^{-1})	(ii) Unknown ion (cm^{-1})
5500.8	1 719 865
4616.5	1 706 927
3808.2	1 685 468
2467.8	1 645 965

   (a) Identify the principal quantum numbers (n) for the upper and lower states for each of the transitions listed in (i), given that these terminate on a common lower level, and determine a value for the Rydberg constant $\mathcal{R}_H$ of atomic hydrogen. [Hint: one line is missing from column (i)]

   (b) Suggest an identification of the unknown ion and assign the transitions in (ii). Derive a value of the Rydberg constant for this ion, $\mathcal{R}_X$, and verify that the ratio $\mathcal{R}_X/\mathcal{R}_H$ is in agreement with that predicted from the reduced masses on the basis of your identification ($m_{e^-} = 5.486 \times 10^{-4}$ u).

2. (a) Predict the number of fine-structure components that would be observed for the $n = 4 \rightarrow 3$ emission line of hydrogen, if the effects of spin–orbit coupling could be fully resolved in the spectrum, and draw a Grottrian diagram to illustrate these transitions.

   (b) One of the $n = 5$ states of hydrogen is split by spin–orbit coupling into two levels with an energy difference of 0.0039 cm^{-1}. Determine the l quantum number for this state and predict the analogous splitting for Li^{2+} (the fine-structure constant $\alpha = 0.007\,297\,3$).

3 The spectra of the alkali metals

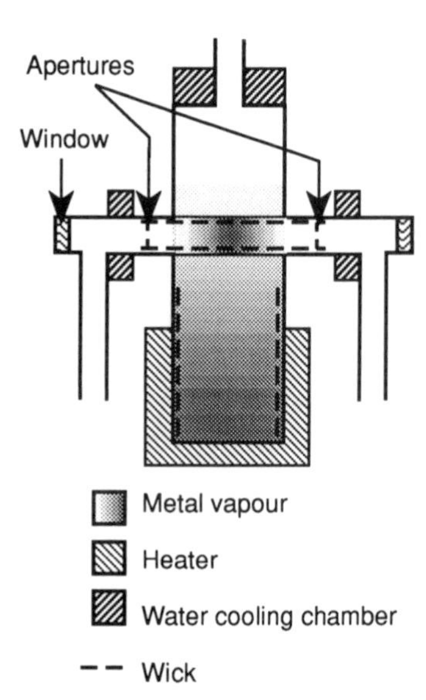

Fig. 3.1 A heat pipe for absorption studies of alkali metals. The metal vapour is primarily confined within the region between the two apertures, and water-cooling beyond the apertures condenses metal vapour onto the tube walls rather than the windows (through which the light is transmitted).

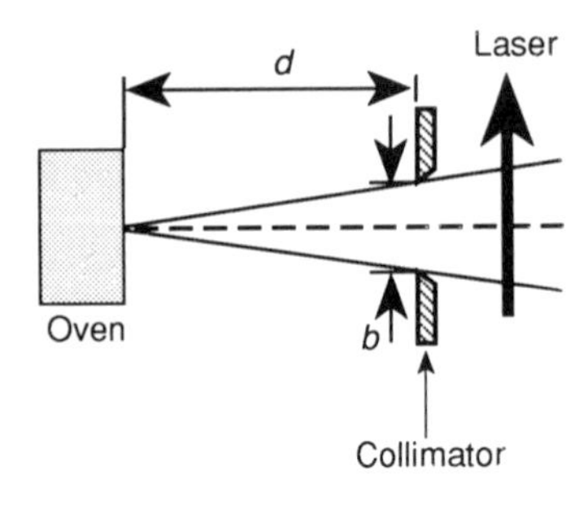

Fig. 3.2 An atomic beam spectroscopy arrangement. The Doppler frequency spread in the perpendicular direction decreases as d/b increases.

3.1 Practical spectroscopy

The alkali metals are solids at room temperature, but a number of methods are available to obtain atoms in the vapour phase. For the recording of an absorption spectrum, the simplest method is to heat the solid sample until a sufficient vapour pressure is obtained. As the melting points and boiling points of the alkalis are quite low (m.p. 180.5–28.4 °C, b.p. 1342–669.3 °C) the degree of heating required is relatively small. Some care has to be exercised to avoid condensation of solid metal on the windows of the sample cell, and a suitable arrangement is the heat pipe illustrated in Fig. 3.1.

For high-resolution studies of the alkali metal absorption spectra, *atomic beams* have been used. An atomic beam is a unidirectional stream of atoms which is formed by allowing gas at high pressure to expand through a small hole into a low-pressure vacuum chamber, with some apertures in the path of expansion to collimate the gas flow. An arrangement for forming a sodium atom beam is shown in Fig. 3.2; the sodium is heated in an oven, and the gas escapes though an aperture of typical diameter 0.1 mm. One of the advantages of atomic beams is that the velocity spread of the atoms in the direction perpendicular to the beam flow is substantially reduced from the normal thermal velocity spread. Consequently the contribution to the linewidth of the observed transitions from the Doppler effect (see Sections 1.2, 2.12) is substantially reduced.

For emission spectroscopy, a suitable pressure of excited atoms can be produced in the hollow cathode lamp shown in Fig. 3.3. The cathode is coated with the metal whose spectrum is required, and the impact of ions on the electrode leads to ejection of atoms from the surface of the metal into the gas phase. Further collisions with fast electrons and ions leads to excitation of the atoms, which then decay by emission.

3.2 The energy levels of many-electron atoms

The energy levels of many-electron atoms will inevitably have a more complicated structure compared to hydrogen, because the quantized energies are now determined not only by electron–nucleus attraction, but also by the mutual repulsion between the electrons. An energy level for the many-electron case represents the *total* energy of all N electrons when the atom is in a particular quantum state. The wavefunctions of the atom Ψ associated with each energy level are now multidimensional functions of $3N$ coordinates ($r_1, \theta_1, \phi_1, r_2, \theta_2, \phi_2, r_3, \ldots$, where r_i, θ_i, ϕ_i are the spherical polar coordinates of electron i). In quantum mechanics the wavefunctions and energy levels are obtained by solving the Schrödinger equation, which for a many-electron atom can be written down in the following form:

$$\left[\sum_i \frac{-\hbar^2}{2\mu}\nabla_i^2 + \sum_i V_{iN} + \sum_{i\neq j} V_{ij}\right]\Psi = E\Psi. \tag{3.1}$$

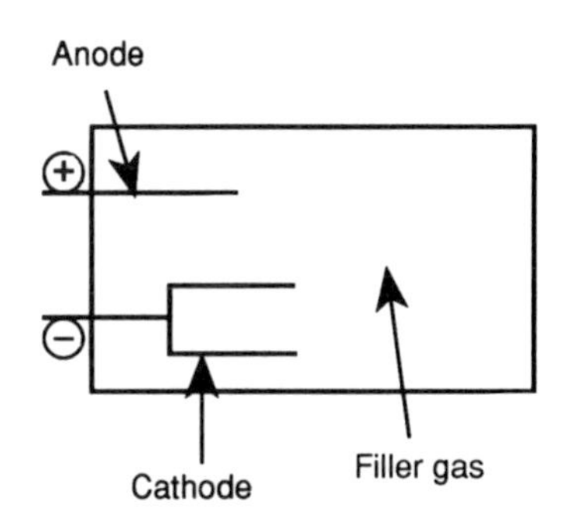

Fig. 3.3 Simplified illustration of a hollow cathode discharge tube.

The first term in square brackets is the sum of kinetic energy operators for each electron, while the second is the sum of potential energies of electron–nucleus attraction. Whereas both these terms are present for a one-electron atom, the new term which must be added is the electron–electron repulsive potential energy, V_{ij} summed over all pairs of electrons $(i \neq j)$.

$$V_{iN} = \frac{-Ze^2}{4\pi\epsilon_0 r_i} \qquad V_{ij} = \frac{e^2}{4\pi\epsilon_0 \mid \mathbf{r}_i - \mathbf{r}_j \mid}. \tag{3.2}$$

Unfortunately, the consequence of this third term is that the equation can no longer be solved exactly. For a two-electron atom such as helium, we have a so-called 'three-body problem' for which even in classical mechanics there is no analytical solution to the equations of motion of the three interacting particles.

To make any progress towards understanding the underlying physics of a many-electron atom, the normally adopted procedure is to make the *orbital approximation*. First, it is assumed that the total wavefunction describing the positions and momenta of all the electrons can be written as a product of one-electron wavefunctions, each of which describes the position and momenta of just one electron.

$$\Psi = \phi_a(1)\phi_b(2)\phi_c(3)\ldots, \tag{3.3}$$

where $1, 2, 3$ label the electrons, and a, b, c label different three-dimensional one-electron wavefunctions. (As we shall see later, this has to be written in a more complicated form to obey the Pauli principle.) It is then proposed that each one-electron wavefunction is similar to the orbitals we have already described for the hydrogen atom, and is characterized by the same set of four quantum numbers n, l, m_l and m_s. Within the orbital approximation each electron moves in the time-averaged charge distribution due to all the other electrons, and the one-electron wavefunctions are obtained by solving a one-electron Schrödinger equation,(3.4), with the potential energy in a form that accounts for the average interactions with all other electrons.

$$\left[\frac{-\hbar^2}{2\mu}\nabla_i^2 + V_{iN} + \overline{\overline{\sum_{i\neq j} V_{ij}}}\right]\phi_i = \epsilon_i\phi_i. \tag{3.4}$$

The double bar on the electron–electron repulsion operator indicates two averaging processes; an averaging over all angles, so that the angle-dependent potential energy is replaced by a spherically symmetric one, and also the averaging over all positions of the other electrons. The only variables in the equation are now the coordinates of electron i and the equation is solvable by numerical integration methods. The complication is that in order to obtain the wavefunction for one electron we need to know the wavefunctions for all the other electrons, so that the average electron–electron repulsion potential can be calculated; these, however, can only be determined by the same procedure which requires a predetermination of the wavefunction for the first electron.

A useful procedure to solve this problem is the *self-consistent field (SCF)* method, illustrated in Fig. 3.4, in which an initial guess is made for the wavefunctions of the $N - 1$ electrons and the Schrödinger equation solved for the Nth electron. This wavefunction is then used as an improved guess for that electron in the calculation of the wavefunction for another electron. The procedure continues iteratively, improving each wavefunction a succession of times until further iterations lead to no further change of the wavefunctions. The total wavefunction is then *self-consistent* and is assumed to be the best solution. The *orbital energy* is defined as the SCF one-electron energy calculated as described above. The total energy of all the electrons is not exactly equal to the sum of orbital energies because that includes the repulsion between each pair of electrons twice over—once when calculating the energy of electron i in the field of electron j, and once in calculating the energy of j in the field of i. Therefore, the total energy is actually

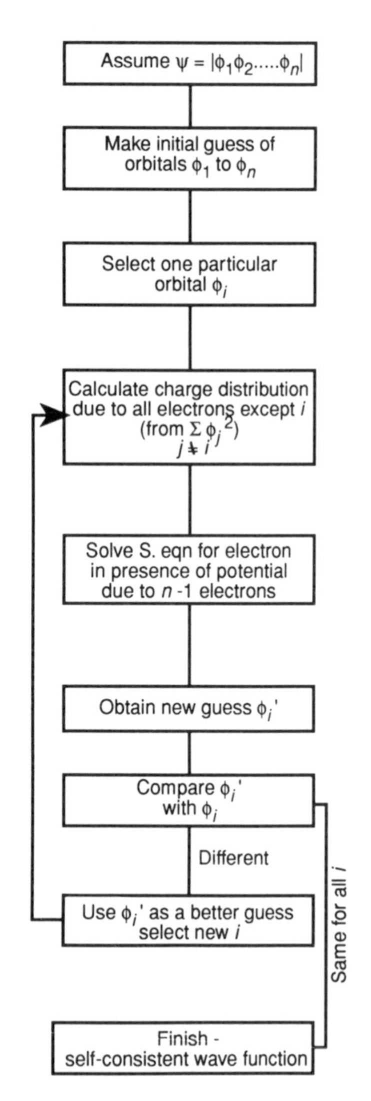

Fig. 3.4 Flow diagram illustrating the self-consistent field procedure.

$$E_{\text{total}} = \sum_i \epsilon_i - V_{\text{ee}} \qquad V_{\text{ee}} = \left\langle \sum_{i\neq j} V_{ij} \right\rangle, \tag{3.5}$$

where the ϵ_i are the orbital energies for each electron and V_{ee} is the expectation value of the total electron–electron repulsion energy.

It should be noted that although the one-electron functions are generally described by the same set of quantum numbers as the hydrogenic orbitals, the radial distribution functions for each orbital may be somewhat different from those of hydrogen, although they will still have the same qualititative appearance, i.e., the same number of radial nodes. On the other hand the l and m_l quantum numbers represent spherical harmonic functions that are *identical* to those of the hydrogen atom.

3.3 Electron configuration and the Pauli exclusion principle

The *electron configuration* is the description of which orbitals are occupied by the electrons in the atom. For example, the helium atom in its state of lowest energy, known as the *ground state* has the configuration $1s^2$ implying both electrons are occupying the 1s orbital. For sodium the ground state electron configuration is $1s^2\,2s^2\,2p^6\,3s$. If the electrons were completely independent, the lowest energy state available to each electron would be that associated with the $1s$ wavefunction and the sodium atom would be expected to have the configuration $1s^{11}$. This is not the case because electrons in atoms are found to obey the *Pauli exclusion principle*, which states that no two electrons in an atom can have the same set of four quantum numbers n, l, m_l and m_s. Thus, the $1s$ orbital has quantum numbers $n = 1, l = 0$ and $m_l = 0$, and there are two possible values for m_s $(=\pm\frac{1}{2})$; therefore, two electrons can exist in this orbital in one atom, one with $m_s = +\frac{1}{2}$, and one with $m_s = -\frac{1}{2}$. For the $2p$ orbital, $n = 2$ $l = 1$ and $m_l = +1, 0$ or -1. The total number of combinations of m_l and m_s is six ($m_l = +1, m_s = \pm\frac{1}{2}$; $m_l = 0, m_s = \pm\frac{1}{2}$; $m_l = -1, m_s = \pm\frac{1}{2}$). Thus six electrons in an atom can occupy a $2p$ orbital. We refer to the set of orbitals of given principal quantum number as a *shell*. Sometimes the nomenclature K shell, L shell ... is used to represent the $n = 1, 2, \ldots$ sets of orbitals. The set of orbitals having common values of both n and l are called a *subshell*, e.g., the $2p$ subshell, $3d$ subshell etc.

To obtain the ground state electron configuration of any atom, the Pauli exclusion principle is applied in conjunction with the *Aufbau principle*; the electrons are added to the orbitals in order of increasing energy, normally $1s, 2s, 2p, 3s, 3p, 4s, 3d, \ldots$ (see Section 3.5 for a discussion of the ordering). Each nl subshell is filled up (two electrons for s, six electrons for p and ten for d) before electrons are added to the next one. When degenerate orbitals are available (e.g., the three $2p$ orbitals), the electrons fill up the degenerate orbitals such that the number of unpaired electrons is maximized. Thus, for example, the silicon atom has 14 electrons, and following the Aufbau principle gives the ground state configuration $1s^2\,2s^2\,2p^6\,3s^2\,3p^2$, with the two $3p$ electrons having different m_l quantum numbers. This Aufbau principle is subject to a number of exceptions as will be discussed later.

3.4 Spectra of many-electron atoms

Nearly all absorption or emission transitions observed in the spectra of many-electron atoms can be described as a change of electron configuration of the atoms, in which just *one* electron changes its orbital. For example, the absorption processes observed in the visible spectrum of the sodium atom can be described as

$$1s^2\,2s^2\,2p^6\,3s \rightarrow 1s^2\,2s^2\,2p^6\,np, \qquad (3.6)$$

with $n = 3, 4, 5 \ldots$. Note that describing a spectroscopic transition in such a format automatically implies the validity of the orbital approximation. As we shall see later, this approximation is not universally valid, but is normally sufficiently close to the truth to be of great value. The change of configuration is governed by the same selection rules as for the hydrogen atom, namely $\Delta l = \pm 1$ and Δn has no restriction. Before considering such processes, we must consider the effects of electron–electron repulsion on the orbitals and their energies in more detail.

3.5 Penetration and shielding

The energy of an electron in a many-electron atom depends on the opposing effects of attraction to the positively-charged nucleus and repulsion by the other electrons. Consider a lithium atom in the excited configuration $1s^2\,3d$. The radial probability distributions for the $1s$ and $3d$ electrons are shown in Fig. 3.5 (assuming these are identical to the hydrogen atom). It can be seen that the probability is very high that the $3d$ electron will be found outside the $1s^2$ core electron distribution. Simple electrostatics shows that the potential of

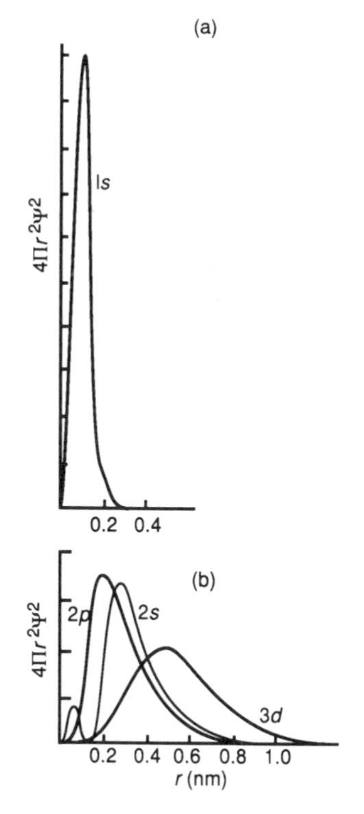

Fig. 3.5 Radial distribution functions for (a) $1s$ and (b) $2s$, $2p$ and $3d$ hydrogenic orbitals.

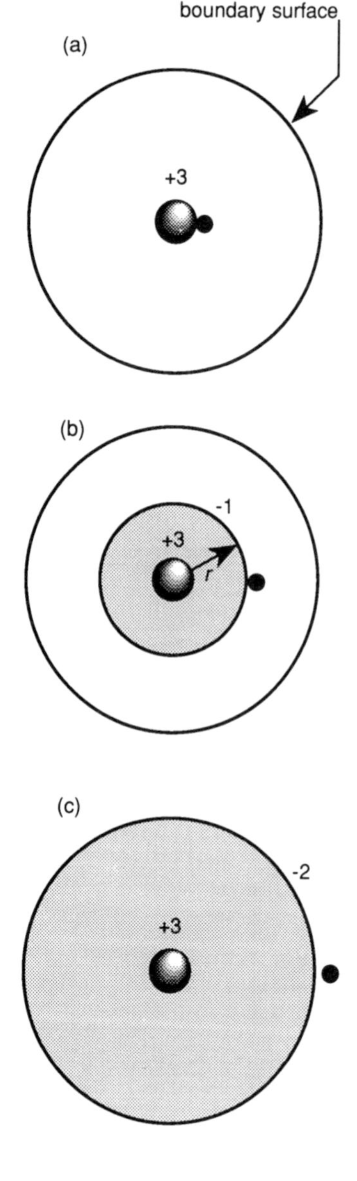

Fig. 3.6 (a) The electron experiences the full nuclear charge (completely inside core electron distribution). (b) The nuclear charge is shielded by that part of the electron distribution inside the sphere of radius r.
(c) The nuclear charge is shielded by all core electrons.

an electron *outside* a hollow charged sphere is equal to the potential due to a point charge at the centre of the sphere (Fig. 3.6c). The $3d$ electron therefore experiences an attractive force due to the positive nuclear charge of $+3$, but a repulsive force due to an effective point charge of -2 arising from the two $1s$ electrons. The net result is that the $3d$ electron experiences an overall effective nuclear charge, $Z_{\rm eff} = +1$: we say that the electron is *shielded* from the nuclear charge by the inner electrons and the binding energy of this electron (i.e., the energy required to remove it from that orbital to infinity) is almost identical to the binding energy of an electron in the same orbital in the hydrogen atom.

Conversely, the two $1s$ electrons are nearly always completely inside the $3d$ electron distribution; electrostatic theory shows that a charged particle *inside* a hollow charged sphere experiences no net force (Fig. 3.6a). Consequently, the two $1s$ electrons experience a much higher effective nuclear charge, $Z_{\rm eff} > 2$ and their binding energy is much greater than in the hydrogen atom. In fact, the stronger attraction to the nucleus will draw the $1s$ orbital radial probability distribution much closer to the nucleus compared to the hydrogen atom.

Consider next the configurations $1s^2\,2s$ (the ground state) and $1s^2\,2p$; although in both cases the $n = 2$ electron spends most of its time outside the $1s^2$ core electron distribution, there is a certain probability of the electron *penetrating* inside the core (see Fig. 3.5); this probability is substantially greater for the $2s$ compared to the $2p$ orbital. When completely inside the $1s$ core the electron would experience a full nuclear charge of $+3$, but when in the outer region, electron repulsion shields the electron resulting in an effective nuclear charge of $+1$. On average it is found that the effective nuclear charge is 1.28 for the $2s$ and 1.02 for the $2p$ electrons, and therefore the $2s$ electron is bound more tightly than the $2p$, i.e., the $2s$ electron is more penetrating than the $2p$ electron. For both the $2s$ and $2p$ orbitals the radial distribution will be drawn in closer to the nucleus, compared to the same orbitals in hydrogen, more so for the $2s$ orbital.

The *non-degeneracy* of orbitals of the same n but different l is a major departure from the H-atom situation. The total energy of the $1s^2\,2s$ configuration is lower in lithium compared to the $1s^2\,2p$ configuration, and in both cases the energy required to remove the outer electron to infinity is greater than the equivalent binding energy in hydrogen as shown in Table 3.1.

Arguments of the type described above can be used to explain the relative energies of the various configurations in almost any atom. In general the penetrating behaviour of an electron decreases with increasing l; the centrifugal force throws the electron away from the nucleus as l increases; therefore, the orbital energies follow the general order,

$$ns < np < nd < nf. \tag{3.7}$$

3.6 The spectra of the alkali metals

The alkali metal atoms in their ground state configurations have a single electron in an ns orbital and a core of inner electrons of lower n (see Table 3.2). The inner core has an overall spherical charge distribution, and for the most part the outer electron is likely to be found outside this core. It is therefore to

Table 3.1
Ionization energies from orbitals in hydrogen and lithium

Atom	Orbital	Ionization energy (cm^{-1})
H	$2s$	27 419.5
H	$2p$	27 419.5
Li	$2s$	43 487.2
Li	$2p$	28 583.4

Table 3.2
Ground state configurations of the alkali metals

Atom	Configuration
Li	$1s^2 2s$
Na	$1s^2 2s^2 2p^6 3s$
K	$1s^2 2s^2 2p^6 3s^2 3p^6 4s$
Rb	$1s^2 2s^2 2p^6 3s^2 3p^6 3d^{10} 4s^2 4p^6 5s$

be expected that in its spectral behaviour an alkali metal atom will behave as a pseudo one-electron atom, similar to hydrogen, but subject to some modifications resulting from penetration and shielding effects. The outermost electron is very different energetically and spatially from the core electrons and the observed spectroscopic transitions involve movement of this outer electron between different orbitals. (Transitions involving the inner orbitals only occur in the X-ray region.) In the following discussion we will drop the core electrons from the electron configuration, but the presence of this core is implicit.

A Grotrian diagram showing the energy levels of the sodium atom is illustrated in Fig. 3.7, and we note the following points.

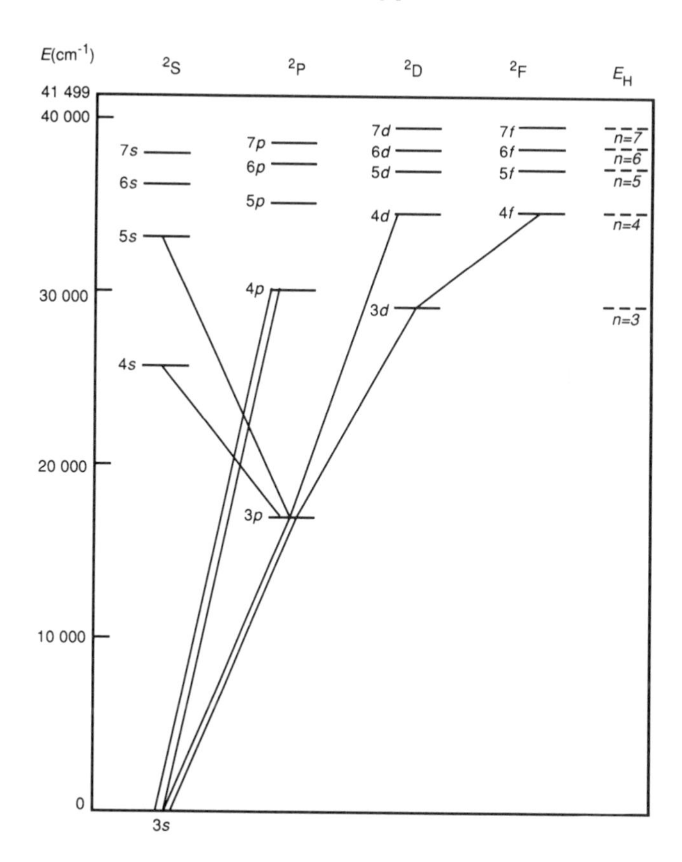

Fig. 3.7 Grotrian diagram for the sodium atom. The double lines indicate transitions observed in both absorption and emission. On the far right-hand side, some of the hydrogen atom energy levels, relative to the ionization limit, are drawn on the same scale.

1. The non-degeneracy of levels of different l but the same n is shown in the figure; it means that the lowest transition in absorption from the ground state is $3p \leftarrow 3s$; in the hydrogen atom, there is only a very small energy gap, between these levels, and such a transition is not considered a part of the normal spectrum. The effects of penetration are so dramatic that the $(n+1)s$ levels actually lie lower in energy than the nd levels.

2. The selection rules $\Delta l = \pm 1$, $\Delta j = 0, \pm 1$ and no restriction for Δn are obeyed as before; hence, in absorption from the ground state, the only transitions observed are $np \leftarrow 3s$, $n = 3, 4, \ldots$, known as the principal series. In emission a number of different series are observed; e.g.,

$ns \rightarrow 3p$	sharp series
$np \rightarrow 3s$	principal series
$nd \rightarrow 3p$	diffuse series
$nf \rightarrow 3d$	fundamental series.

As a result of the different energies of the s, p and d states these transitions are not almost degenerate, as in the hydrogen atom, but are spectrally distinct—this is shown by the emission spectrum in Fig. 3.8 which shows two of these series (sharp and diffuse). Note that the lettering s, p, d, f for the orbital angular momentum states is historically derived from the names of these series.

3. Spin–orbit coupling is again evident as shown by the double lines in the spectrum of Fig. 3.8, and the same states are obtained as for the hydrogen

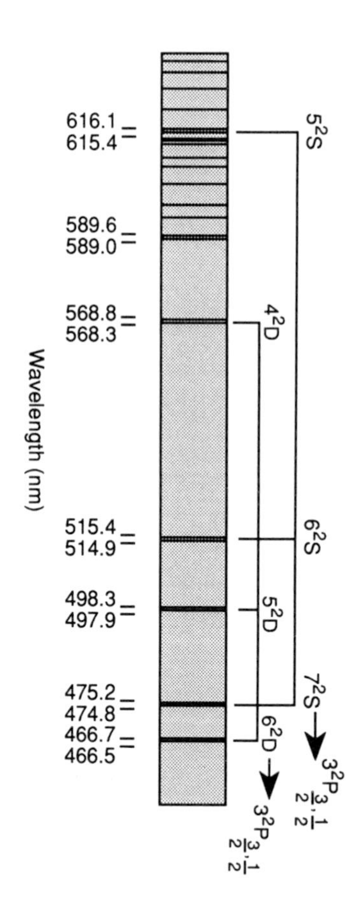

Fig. 3.8 Photographic plate emission spectrum of the sodium atom. The number preceeding the term symbol ^2S, ^2D or ^2P is the principal quantum number of the outer electron. The doubling of lines is due to spin–orbit coupling.

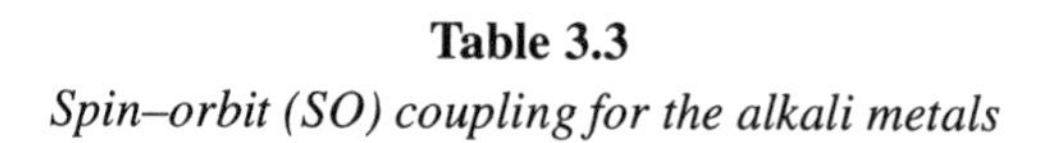

Table 3.3

Spin–orbit (SO) coupling for the alkali metals

Atom	Orbital	Z^4(10^3 a.u.)	SO coupling constant (cm^{-1})
Li	$2p$	0.08	0.3
Na	$3p$	14.64	17.2
K	$4p$	130.3	57.7
Rb	$5p$	1874	237.6
Cs	$6p$	9150	554.1

atom, i.e., following the rule $j = l + s, \ldots \mid l - s \mid$, $(= l \pm \frac{1}{2}$, for $s = \frac{1}{2})$, leads to the levels

$3s$	$^2S_{\frac{1}{2}}$ (no splitting)
$3p$	$^2P_{\frac{1}{2}}$ and $^2P_{\frac{3}{2}}$
$3d$	$^2D_{\frac{3}{2}}$ and $^2D_{\frac{5}{2}}$.

The selection rule $\Delta j = 0 \pm 1$ permits three components for each $nd \rightarrow 3p$ transition, but the $^2D_{\frac{3}{2}} \rightarrow {}^2P_{\frac{3}{2}}$ and the $^2D_{\frac{5}{2}} \rightarrow {}^2P_{\frac{3}{2}}$ components are not separated at the level of instrumental resolution in Fig. 3.8. The spin–orbit splitting is much larger than in the hydrogen atom as shown by the splittings of the lowest p states in Table 3.3. The alkali metal atoms do not show a Z^4 dependence of the splitting as in the hydrogenic atoms—the formulae derived in Section 2.11 were not appropriate for the many-electron case. Nevertheless, there is a general increase in splitting with increasing nuclear charge. The formula (2.62) expressing the dependence of the spin–orbit energy on j can be applied in the case of the alkali metals and some other many-electron atoms, but the coupling constant ζ is normally replaced by the symbol A, and this constant is usually determined by experiment.

3.7 The quantum defect

As stated in Section 3.2, the energy levels of a many-electron atom include the summed energies of all N electrons. However, in general, we are not so much interested in the total energy, as in the *relative* energy of various configurations. For the configurations of the alkali metals, it is convenient to express the energy levels relative to the ground state energy level as shown in Fig. 3.7. For the hydrogen atom, the energy levels relative to the ground state, E', are given by the formula

$$E' = I - \frac{\mathcal{R}hc}{n^2}, \tag{3.8}$$

where I is the ionization energy of a $1s$ electron ($= \mathcal{R}hc$). However, for the non-hydrogenic atom, the simple Rydberg formula is no longer valid even if spin–orbit coupling is ignored; in general, the energies are lower than predicted by eqn (3.8), indicating that an electron of given n, l quantum numbers is more strongly bound in a many-electron atom than in a hydrogen atom; the effective nuclear charge experienced by the outer electron is greater than one (see Section 3.5). One way to deal with the situation is to define an *effective principal quantum number* ν_{nl} for energy level E'_{nl}, such that

$$E'_{nl} = I - \frac{\mathcal{R}hc}{\nu_{nl}^2}. \tag{3.9}$$

The effective principal quantum number is essentially an experimental quantity, and will be less than the actual principal quantum number. The quantum defect δ_{nl} is then defined by the equation $\nu_{nl} = n - \delta_{nl}$ and eqn (3.9) is then rewritten as

$$E'_{nl} = I - \frac{\mathcal{R}hc}{(n - \delta_{nl})^2}. \tag{3.10}$$

The quantum defect can be considered as a measure of the degree of penetration of the outermost electron.

It was originally an empirical observation that the quantum defects for levels of different n but the same l in a given atom have a value that is approximately constant. Table 3.4 lists the observed average quantum defects for the ns, np and nd levels in sodium. A theoretical justification of this finding is possible via 'quantum defect theory'. Qualitatively, it may be considered that when the electron penetrates close to the nucleus, its kinetic and potential energies have (temporarily) very large magnitudes. The difference of the magnitudes of these two quantities gives the total energy, which is very small in comparison with the local potential energy. A change in principal quantum number (a change in the total energy) has a very small fractional effect on the local kinetic energy and therefore the wavefunction in the inner region is almost independent of n, giving rise to the constant quantum defect.

Table 3.4
Quantum defects for sodium

Orbital	Quantum defect
ns	1.35
np	0.85
nd	0.01

3.8 Transition energies and the determination of ionization energies

The transition energies in the spectra of the alkali metals are given by

$$\Delta E(nl \to n'l') = -\mathcal{R}hc\left[\frac{1}{(n-\delta_{nl})^2} - \frac{1}{(n'-\delta_{n'l'})^2}\right]. \tag{3.11}$$

The ionization energy is defined, as before, as the energy required to remove the outermost electron to infinity and so

$$I = \frac{\mathcal{R}hc}{(n-\delta_{ns})^2}, \tag{3.12}$$

where n is 2 for Li, 3 for Na, etc.

A useful method for determining the ionization energy is to use an extrapolation procedure for the transition energies of an emission series for which the upper state quantum defect is negligible and spin–orbit splitting small. The diffuse series serves this purpose well; ignoring the nd quantum defect, the energies of the transitions are given by

$$\Delta E(nd \to n'p) = -\mathcal{R}hc\left[\frac{1}{n^2} - \frac{1}{(n'-\delta_{n'p})^2}\right]. \tag{3.13}$$

The second term in the square bracket is a constant for a given emission series, and therefore, a plot of ΔE versus $1/n^2$ gives the binding energy of an electron in an $n'p$ orbital, $I(n'p) = -\mathcal{R}hc/(n'-\delta_{n'p})^2$ as the intercept. The ionization energy from the ground state is then obtained by adding the observed $n'p \to ns$ transition energy. Note that for the most careful extrapolation, we should use the transition energies for just one of the two $n'p$ spin–orbit components. In a more general case, where the series involving zero-quantum-defect states are not available, iterative procedures must be adopted to determine the ionization energy. An example of such a procedure is provided as a worked example in problem 2 (section 3.10); in practice, a high-speed computer would be used to perform a nonlinear least squares fit of the data to eqn (3.11).

3.9 Laser cooling of sodium atoms

An exciting new application of atomic spectroscopy makes use of lasers to slow down atoms in the gas phase to such low velocities that their kinetic energy distribution is characteristic of that expected for a gas at temperatures of order 10^{-6} K; in the ultimate experiments, submicro-Kelvin temperatures have been possible. In this section some experiments of this type performed on sodium atoms are illustrated.

The method makes use of the linear momentum associated with a photon which is calculated from the de Broglie relationship

$$p = \frac{h}{\lambda}. \tag{3.14}$$

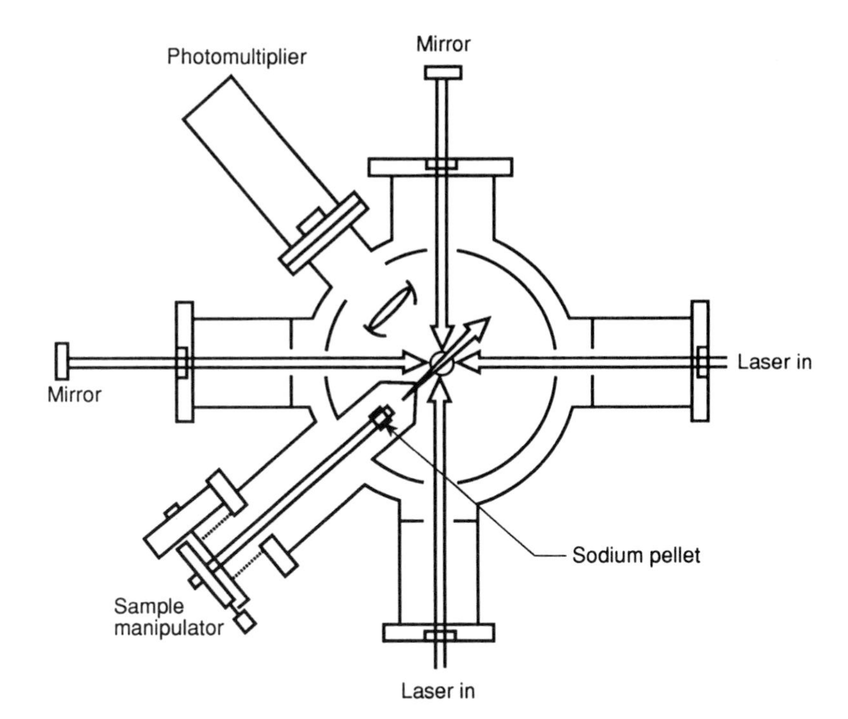

Fig. 3.9 Laser-cooling experimental set-up. A low partial pressure of sodium atoms leaks into the central chamber from the sample manipulator, where it interacts with six perpendicular laser beams (four shown and two perpendicular to the page). A photomultiplier is used to detect fluorescence from the laser-cooled atoms.

If a photon is absorbed by an atom moving towards it, then conservation of linear momentum dictates that the momentum of the atom must decrease. In the set-up illustrated in Fig. 3.9, six laser beams are directed at a sample of the atoms; the beams, pointing in the directions $+x$, $-x$, $+y$, $-y$, $+z$, $-z$, are all tuned to a frequency just below that required to induce a transition in the atoms to be cooled. The photons will then only be absorbed if the Doppler effect is such as to shift the effective frequency upwards, which only occurs if the atom and photons are moving towards each other. All six beams slow down those atoms moving towards them and therefore the average velocity components of the atoms are decreased in all directions. By slowly moving the laser frequency closer to resonance, the maximum velocity component in all directions is decreased.

One of the most convenient atoms for study in this way is the sodium atom, for which the strongly allowed transition $3p \leftarrow 3s$ lies in the visible region where frequency-tunable lasers are available. Sufficient sodium atoms are present in the vapour pressure above metallic sodium in an evacuated chamber to perform the experiments using the experimental arrangement indicated. The use of a magnetic field can assist the cooling and trapping of the atoms, through the Zeeman effect discussed in Chapter 5. More details of the experiments can be found in the Background reading. The velocities of the atoms can be slowed down to such an extent that they can be observed to fall under the action of gravity. In effect the cloud of cooled atoms can be considered as a unique example of a fully quantized system; the temperature is so low that under some circumstances even the quantization of translational energy states becomes important; there are consequently valuable opportunities to test important laws of physics under conditions of fundamental simplicity.

3.10 Problems

1. The *fundamental* series of the emission spectrum of sodium shows transitions at the vacuum wavelengths (nm)

$$1846.4, 1268.27, 1083.75, 996.38.$$

Assuming the quantum defects for the nf levels are zero, and using the $3s \rightarrow 3d$ separation of 29 172.9 cm^{-1} (ignore the spin–orbit splitting of the $3d$ level), estimate the ionization energy of sodium ($\mathcal{R}_{Na} = 109\,735$ cm^{-1}).

2. For the *principal* series of potassium, four sequential lines occur at the wavelengths (nm)

$$321.833, 310.279, 303.573, 299.311.$$

Determine the ionization energy of potassium using an iterative procedure as follows;

(a) Make a rough guess for the ionization energy I_{trial} (the wavenumber to which you think the series converges).

(b) Using I_{trial} calculate the effective principal quantum number (ν_1) of the first line at 321.833 nm and compare it with that calculated for the fourth line at 299.311 nm (ν_4). If your guess for I_{trial} is accurate, then these values should differ by three units ($\mathcal{R} = 109\,736$ cm^{-1}).

(c) If not, use $\nu_1 + 3$ as an improved estimate for ν_4 and then use this value and the transition wavenumber for line 4 to recalculate the ionization energy; ν_1 is less sensitive to errors in I_{trial} than ν_4, but the transition wavenumber for line 4 can be used as a better predictor of the ionization energy.

(d) Go back to (b) and repeat until successive iterations make no change to the ionization energy (to an accuracy of 2 cm^{-1}). At this point the effective quantum numbers should differ by 1.00 between each line (to an accuracy of two decimal places).

4 The spectrum of the helium atom

To this point we have discussed the spectra of those atoms that can be considered as one-electron species. In these cases the motion of the outermost electron is essentially independent of that of the inner electrons, except that it experiences a repulsive interaction with the average charge distribution of the inner electron core (if present). The magnetic moments associated with the orbital and spin angular momenta of the core electrons are cancelled out within the closed shell, and therefore only the outer electron angular momentum is important.

The excited states of the helium atom cannot be considered in this manner, as there are now two electrons in unfilled shells, and the interactions between the electrons have to be taken into account more explicitly. There is now more than one source of angular momentum in the atom, both electrons having spin angular momentum and possibly both having orbital angular momentum. The coupling of these angular momenta has important effects on the energy levels. The degeneracies of the various configurations are generally higher than for the one-electron case, because of the many combinations of the m_l and m_s quantum numbers for the two electrons. However, it is shown below that electron-electron interactions lead to a lifting of these degeneracies, and hence to a complication of the energy level structure compared to hydrogen or the alkali metals.

4.1 Total spin angular momentum—singlets and triplets

Consider first the ground state configuration in the helium atom $1s^2$. As both electrons have the same set of quantum numbers, n, l and m_l, the Pauli exclusion principle requires them to have different m_s quantum numbers, $\pm\frac{1}{2}$. The electrons are *spin-paired* and the resultant magnetic moment is zero. Such a state is referred to as a *singlet*. In contrast, for the excited state configurations of the type $1s\,nl$ (of which the lowest energy is $1s\,2s$) there are now no restrictions from the Pauli principle on the m_s quantum numbers; therefore, there are two possibilities for the relative spin orientations as shown in Fig. 4.1, either parallel or opposed (i.e., the two m_s quantum numbers are the same or different). These possibilities are different quantum states of the atom, and will, in general, have different energies. In the former case, there will be a resultant magnetic moment due to the electron spins, and this is referred to as the *triplet* state, while the singlet state is again one in which there is no resultant magnetism.

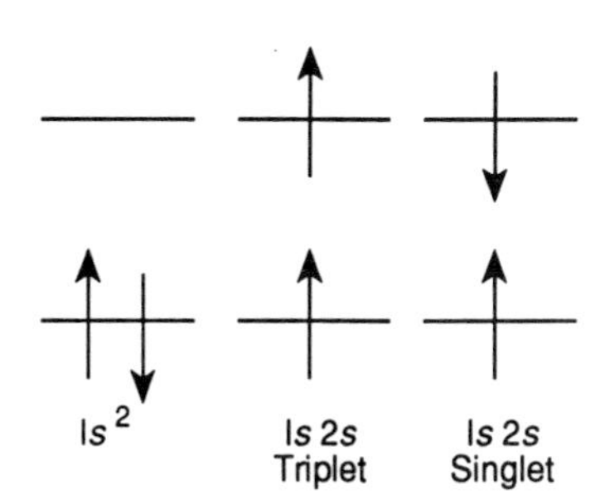

Fig. 4.1 A simple representation of parallel and opposed spin orientations for $1s^2$ and $1s2s$ configurations of helium.

The triplet state associated with a particular electron configuration of the atom is always lower in energy than the singlet state associated with the same configuration. Any excited electron configuration of the helium atom is split into at least two energy levels (with the exception of ns^2 states). The reason for the energy difference has little to do with the different magnetic properties of the atom in the two states, but is actually an electrostatic effect known as *spin correlation.* As explained in more detail in Sections 4.4–4.6, the motions of the electrons can no longer be considered as entirely independent of each other. Electron 1 does not merely see the average charge distribution due to electron 2 but is actually affected by the instantaneous position of electron 2 relative to its own position. In a triplet state it can be shown (Section 4.5) that the two electrons are very unlikely to be found in close proximity, whereas in the singlet state there is actually a higher than average probability of the two electrons existing at the same point in space. Consequently there is less electron–electron repulsion in the triplet state than the singlet state and the triplet state has lower energy.

The origin of the 'singlet/triplet' nomenclature becomes apparent if we define the total spin angular momentum vector $\mathbf{S}$ as the vector sum of the spin vectors of the individual electrons, $\mathbf{s}_1$ and $\mathbf{s}_2$.

$$\mathbf{S} = \mathbf{s}_1 + \mathbf{s}_2. \tag{4.1}$$

A total spin angular momentum *quantum number* S is then defined such that

$$|\,\mathbf{S}\,| = \sqrt{S(S+1)}\hbar. \tag{4.2}$$

The total spin angular momentum is quantized in the same manner as the individual spin and orbital angular momenta, and therefore S is restricted to integral values for an even number of electrons or to half-integral values for an odd number of electrons. The quantization, together with eqns (4.1) and (4.2) lead to the conclusion that for $s_1 = \frac{1}{2}$ and $s_2 = \frac{1}{2}$ the possible values for S are either 1 or 0. The state of the atom with $S = 1$ has a resultant magnetic moment and is identified with the triplet state discussed above, whereas the $S = 0$ state is the singlet state. 'Triplet' refers to the *spin multiplicity*; the total spin angular momentum has $2S + 1$ quantized components along the z-direction with $M_S = S, S-1 \ldots -S$; for the triplet state $M_S = +1, 0, -1$ while for the singlet state $M_S = 0$ only.

4.2 The spectrum of helium

Nearly all the observed transitions of the helium atom involve the change of orbital of just one electron. Within the orbital approximation, a photon can only interact with one electron in the atom. In absorption the transitions are from the ground state, and can be represented as

$$1s\,nl \leftarrow 1s^2. \tag{4.3}$$

In emission more generalized processes are possible, formally represented as

$$nl\,n'l' \rightarrow nl\,n''l''. \tag{4.4}$$

The selection rules for change of electron configuration are, as before,

$$\Delta n \text{ no restriction} \tag{4.5}$$

$$\Delta l = \pm 1 \text{ only.} \tag{4.6}$$

A new selection rule, very important for the helium atom, is the total spin selection rule

$$\Delta S = 0. \tag{4.7}$$

The obvious consequence of this rule is that *transitions between singlet and triplet states are forbidden.*

The ground state of helium is a *singlet* state with configuration, $1s^2$, and therefore only the singlet manifold of states are seen in the normal absorption spectrum. Furthermore, because of the $\Delta l = \pm 1$ selection rule, the transitions observed are all of the type

$$1s\,np \leftarrow 1s^2. \tag{4.8}$$

The emission spectrum of helium is observed by dispersing the light from a helium discharge lamp. Ionization of the atoms occurs fairly readily in the discharge and can be followed by ion–electron recombination to form excited helium

$$\mathrm{He}^+ + \mathrm{e}^- \rightarrow \mathrm{He}^*. \tag{4.9}$$

The excited helium atoms He* are then de-excited by emission of radiation. In the recombination event eqn (4.9) the two electrons may end up forming a singlet or a triplet state and therefore transitions are observed between levels within both the singlet and triplet manifolds. Some of the transitions are illustrated in the Grotrian diagram of Fig. 4.2. $1s2s \rightarrow 1s^2$ is forbidden because it breaks the $\Delta l = \pm 1$ rule, as does the $1s3d \rightarrow 1s2s$ transition. $1s2p(^1\mathrm{P}) \rightarrow 1s2s(^3\mathrm{S})$ breaks the $\Delta S = 0$ selection rule, and $1s2s(^3\mathrm{S}) \rightarrow 1s^2(^1\mathrm{S})$ breaks both rules hence is very strongly forbidden. In principle, *doubly-excited* states of helium could also be formed in the discharge (e.g., $2p\,3s$). However, the sum of the two excitation energies is generally greater than the ionization energy of one electron and therefore such states are unstable with respect to a process known as *autoionization* (described in Section 5.11). The atom is very short-lived in these states and preferentially undergoes ionization rather than emission of a photon. Therefore the only important states in the emission spectrum are the singly excited states $1s\,nl$.

It should be noted that for all $1s\,nl$ configurations, the $1s$ electron shields the outer electron from the increased nuclear charge. The extent of shielding depends on the l quantum number, as in the alkali metal atoms; therefore the configurations $1s\,nl$ and $1s\,nl'$ have different energies.

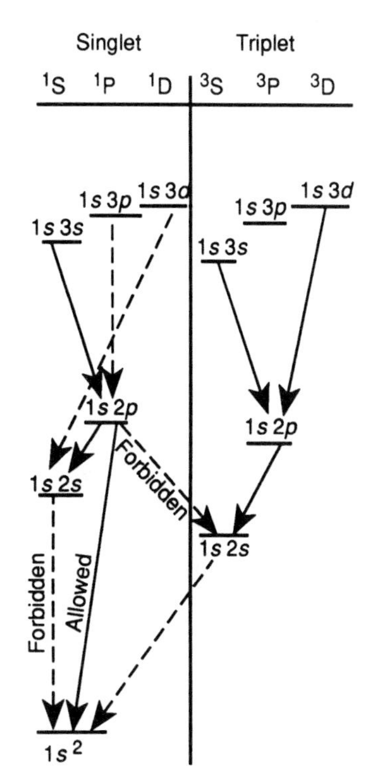

Fig. 4.2 Grotrian diagram showing allowed (bold lines) and forbidden (dashed lines) transitions of helium.

4.3 Spin–orbit coupling

In Section 4.1 a total spin angular momentum vector **S** was defined, which gives rise to a resultant magnetic moment of magnitude $\mathbf{m} = g_e\gamma_e\mathbf{S}$. A total *orbital* angular momentum vector, **L**, can also be defined as the vector sum of the individual electron orbital momenta, $\mathbf{l}_1$ and $\mathbf{l}_2$

$$\mathbf{L} = \mathbf{l}_1 + \mathbf{l}_2. \tag{4.10}$$

For the present we shall only need to consider those states of helium with at least one electron in a $1s$ orbital. Under these circumstances $\mathbf{l}_1 = 0$ and $\mathbf{L} = \mathbf{l}_2$;

the only source of orbital angular momentum is the excited electron. The spin–orbit coupling in the helium atom can be understood quite well by assuming that it is the *resultant* spin that couples with the orbital angular momentum, rather than the individual electron spins. If the state of the atom is singlet, there is no resultant spin magnetic moment and therefore, to a good approximation, no spin–orbit coupling. However, for the triplet configurations of the type $1s\,nl$, where $l \geq 1$, there will be spin–orbit coupling between the resultant spin and orbital angular momenta. By analogy with Sections 2.10 and 3.6, a total angular momentum vector $\mathbf{J} = \mathbf{L} + \mathbf{S}$ is now defined, and the quantum number J can take the values

$$J = L + S, L + S - 1, \ldots \mid L - S \mid . \tag{4.11}$$

For the singlet states $S = 0$, and therefore $J = L$. For the triplet states the possible values of J are

$$J = L + 1, L, L - 1. \tag{4.12}$$

Each J state is $(2J + 1)$-fold degenerate, corresponding to the number of permitted values of M_J the total angular momentum projection quantum number.

For a triplet $1s\,np$ state, $L = 1$ and therefore $J = 2, 1, 0$; for $1s\,nd$, $L = 2$ and $J = 3, 2, 1$. These states are designated by the term symbols ${}^3P_2, {}^3P_1, {}^3P_0$ and ${}^3D_3, {}^3D_2, {}^3D_1$ respectively. The states of the atom with different J have different energies, with the spin–orbit contribution to the energy given by the formula

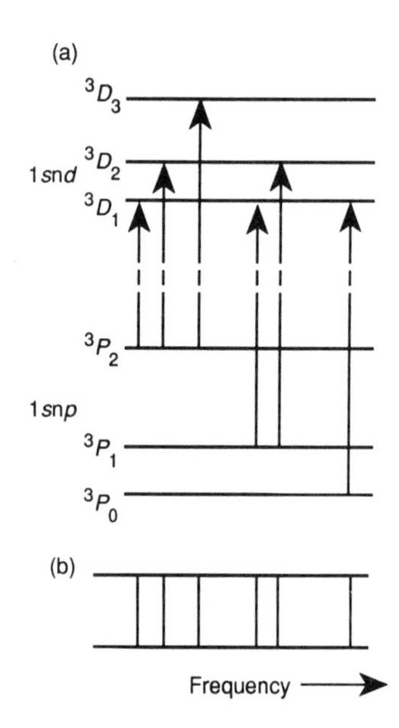

Fig. 4.3 The six permitted components for a $1s\,nd(^3D) \leftarrow 1s\,np(^3P)$ transition with inclusion of spin–orbit coupling; (a) energy levels, (b) spectrum.

$$E_{so} = \frac{1}{2}A[J(J + 1) - L(L + 1) - S(S + 1)]hc. \tag{4.13}$$

This expression for E_{so} is directly analogous to eqn (2.62). However, for helium in its $1s\,np$ triplet states, the spin–orbit coupling constant, A, is very small, and in fact there are other similar interactions which affect the energy levels by a similar or greater order of magnitude. In particular, there is a contribution from the magnetic interaction between the two spin magnetic moments of the two electrons. As a consequence, the Landé interval rule, for the energy difference between two levels (different J values) arising from the same term (same L and S values), which can be straightforwardly derived from eqn (4.13),

$$E(J + 1) - E(J) = (J + 1)Ahc, \tag{4.14}$$

is not obeyed in the helium atom; e.g., the splittings for the $1s2p\ {}^3P$ state are $0.906\ cm^{-1}$ for $J = 1 \rightarrow 0$ and $0.078\ cm^{-1}$ for $J = 2 \rightarrow 1$.

The selection rule for change of J in the spectra is

$$\Delta J = 0, \pm 1, \tag{4.15}$$

except that the transition between two levels with $J = 0$ is not allowed. In Fig. 4.3 the six permitted components for a $1s\,nd(^3D) \leftarrow 1s\,np(^3P)$ transition are illustrated.

4.4 Spin wavefunctions and the Pauli principle

To explain the origin of spin correlation, which leads to the lower energy of triplet states relative to singlet states of the same configuration, it is necessary to consider the total wavefunction for the helium atom in more detail. The total wavefunction for an atom can generally be written as a simple product of a total spin wavefunction Ψ_{spin} and a total orbital wavefunction Ψ_{space}

$$\Psi = \Psi_{\text{spin}}\Psi_{\text{space}}. \tag{4.16}$$

Although Ψ_{space} is obviously a mathematical function of six coordinates in the helium atom, three for electron 1 and three for electron 2, the meaning of Ψ_{spin} is rather less obvious. It will never be necessary (or possible) for us to derive a specific mathematical form for this wavefunction, but it will suffice to introduce the idea that each electron has a fourth coordinate, the spin coordinate. The total spin wavefunction must be a function of the spin coordinates for two electrons, and its (unspecified) form will depend on the m_s quantum numbers of the two electrons, just as the spatial wavefunction depends on the n, l and m_l quantum numbers for each electron.

For the helium atom in its excited states, there are four possible combinations of the m_s quantum numbers of two electrons in different orbitals. It might be assumed that the total spin wavefunction could be written as a simple product of one-electron spin wavefunctions

$$\Psi_1 \equiv \alpha(1)\alpha(2) \tag{4.17}$$
$$\Psi_2 \equiv \beta(1)\beta(2) \tag{4.18}$$
$$\Psi_3 \equiv \alpha(1)\beta(2) \tag{4.19}$$
$$\Psi_4 \equiv \beta(1)\alpha(2), \tag{4.20}$$

where the notation α, β represents the wavefunctions corresponding to $m_s = +\frac{1}{2}$ or $-\frac{1}{2}$ respectively, such that,

$$\alpha(1) \Rightarrow m_s = +\frac{1}{2} \quad \text{for electron 1}$$

$$\beta(2) \Rightarrow m_s = -\frac{1}{2} \quad \text{for electron 2.}$$

However, not all of these wavefunctions in eqns (4.17)–(4.20) are permitted under the *Pauli principle* which is a generalized version of the Pauli exclusion principle discussed in Section 3.3. The Pauli principle states that *the total wavefunction for the atom must be antisymmetric with respect to exchange of electrons.* Using the exchange operator $\hat{P}_{12}$ this principle may be written mathematically as

$$\hat{P}_{12}\Psi = -\Psi. \tag{4.21}$$

The effect of the operator is to exchange the electron labels (1) and (2).

Given the factorization employed in eqn (4.16), it is clear that either the spatial part of the wavefunction must be antisymmetric and the spin part symmetric or vice versa (using the general product rules for functions; symm × anti = anti, symm × symm = symm, and anti × anti = symm). Consider the effect of the permutation operator on the spin wavefunctions given in eqns (4.17) to (4.20).

$$\hat{P}_{12}\,\Psi_1 = \hat{P}_{12}\,\alpha(1)\alpha(2) = \alpha(2)\alpha(1) = +\Psi_1 \tag{4.22}$$

$$\hat{P}_{12}\,\Psi_2 = \hat{P}_{12}\,\beta(1)\beta(2) = \beta(2)\beta(1) = +\Psi_2 \tag{4.23}$$

$$\hat{P}_{12}\,\Psi_3 = \hat{P}_{12}\,\alpha(1)\beta(2) = \alpha(2)\beta(1) = +\Psi_4 \tag{4.24}$$

$$\hat{P}_{12}\,\Psi_4 = \hat{P}_{12}\,\beta(1)\alpha(2) = \beta(2)\alpha(1) = +\Psi_3. \tag{4.25}$$

Ψ_1 and Ψ_2 are *symmetric* with respect to electron exchange, and would be valid functions provided they are combined with an antisymmetric spatial wavefunction. On the other hand, Ψ_3 and Ψ_4 are neither symmetric nor antisymmetric—the effect of $\hat{P}_{12}$ is to produce a different wavefunction—and hence these cannot be valid wavefunctions. However, we can construct linear combinations of these degenerate wavefunctions which are either symmetric or antisymmetic;

$$\Psi^+ \equiv \tfrac{1}{\sqrt{2}}(\Psi_3 + \Psi_4) \equiv \frac{1}{\sqrt{2}}[\alpha(1)\beta(2) + \alpha(2)\beta(1)] \tag{4.26}$$

$$\Psi^- \equiv \tfrac{1}{\sqrt{2}}(\Psi_3 - \Psi_4) \equiv \frac{1}{\sqrt{2}}[\alpha(1)\beta(2) - \alpha(2)\beta(1)]. \tag{4.27}$$

It can be seen that

$$\hat{P}_{12}\Psi^+ = \frac{1}{\sqrt{2}}[\alpha(2)\beta(1) + \alpha(1)\beta(2)] = +\Psi^+ \tag{4.28}$$

$$\hat{P}_{12}\Psi^- = \frac{1}{\sqrt{2}}[\alpha(2)\beta(1) - \alpha(1)\beta(2)] = -\Psi^-. \tag{4.29}$$

The four permitted spin wavefunctions for the helium atom are therefore given by Ψ_1, Ψ_2, Ψ^+ and Ψ^-. Three of these functions are symmetric, while Ψ^- is antisymmetric with respect to the electron exchange operator $\hat{P}_{12}$.

The origin of the Pauli principle lies in the indistinguishability of the electrons; because it is impossible to identify which electron has which set of quantum numbers, the wavefunction must be set up to take this uncertainty into account. If there is one electron with $m_s = -\frac{1}{2}$ and one with $m_s = +\frac{1}{2}$ the wavefunction must show an equal probability for the two possibilities $\alpha(1)\beta(2)$ and $\beta(1)\alpha(2)$.

Each total spin wavefunction must have an associated pair of quantum numbers S and M_S. The M_S quantum numbers for each of these wavefunctions are obtained by summing the individual m_{s_1} and m_{s_2} values, giving $+1$, -1, 0 and 0 respectively, for Ψ_1, Ψ_2, Ψ^+ and Ψ^-. The functions, Ψ_1 and Ψ_2 must undoubtedly be two of the three components of the triplet $S = 1$ state, because the $S = 0$ singlet state has only one component with $M_S = 0$. We will simply state here, without proof, that the third component of the triplet state is the other symmetric spin state, Ψ^+ with Ψ^- being the singlet wavefunction. It should be noted that although the triplet $M_S = 0$ function has no component of spin angular momentum along the z-axis, it *does* have an overall *resultant* spin angular momentum, whereas the singlet $M_S = 0$ function does not. The vector diagrams in Fig. 4.4 illustrate schematically the difference between the two $M_S = 0$ wavefunctions, (a) and (c).

Given that the triplet spin wavefunctions are all symmetric with respect to electron exchange, the orbital part of the wavefunction must be antisymmetric. It is not sufficient to write the orbital wavefunction as a simple product of one-electron wavefunctions, e.g., $\phi_{1s}(\mathbf{r}_1)\phi_{2s}(\mathbf{r}_2)$, as suggested by the orbital approximation, but it must be written as an antisymmetric or symmetric linear combination analogous to those in eqns (4.26)–(4.27); e.g., for the $1s\,2s$ configuration,

$$\Psi^{\pm}_{\text{space}} = \frac{1}{\sqrt{2}}(\phi_{1s}(\mathbf{r}_1)\phi_{2s}(\mathbf{r}_2) \pm \phi_{1s}(\mathbf{r}_2)\phi_{2s}(\mathbf{r}_1)). \tag{4.30}$$

($\mathbf{r}_1$ and $\mathbf{r}_2$ are position vectors from the origin to the position of electrons 1 and 2 respectively, and are a shorthand notation for the full sets of coordinates r_1, θ_1, ϕ_1 etc.) According to the Pauli principle the four valid total wavefunctions, having overall antisymmetric behaviour with respect to electron exchange, for the $1s\,2s$ configuration are

$$\frac{1}{\sqrt{2}}(\phi_{1s}(\mathbf{r}_1)\phi_{2s}(\mathbf{r}_2) - \phi_{1s}(\mathbf{r}_2)\phi_{2s}(\mathbf{r}_1)) \times \alpha(1)\alpha(2) \tag{4.31}$$

$$\frac{1}{\sqrt{2}}(\phi_{1s}(\mathbf{r}_1)\phi_{2s}(\mathbf{r}_2) - \phi_{1s}(\mathbf{r}_2)\phi_{2s}(\mathbf{r}_1)) \times \frac{1}{\sqrt{2}}(\alpha(1)\beta(2) + \beta(1)\alpha(2)) \tag{4.32}$$

$$\frac{1}{\sqrt{2}}(\phi_{1s}(\mathbf{r}_1)\phi_{2s}(\mathbf{r}_2) - \phi_{1s}(\mathbf{r}_2)\phi_{2s}(\mathbf{r}_1)) \times \beta(1)\beta(2) \tag{4.33}$$

$$\frac{1}{\sqrt{2}}(\phi_{1s}(\mathbf{r}_1)\phi_{2s}(\mathbf{r}_2) + \phi_{1s}(\mathbf{r}_2)\phi_{2s}(\mathbf{r}_1)) \times \frac{1}{\sqrt{2}}(\alpha(1)\beta(2) - \beta(1)\alpha(2)). \tag{4.34}$$

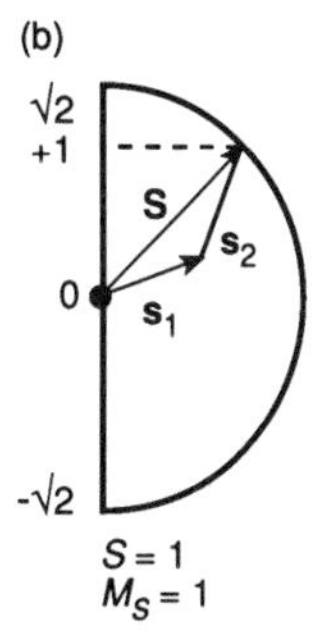

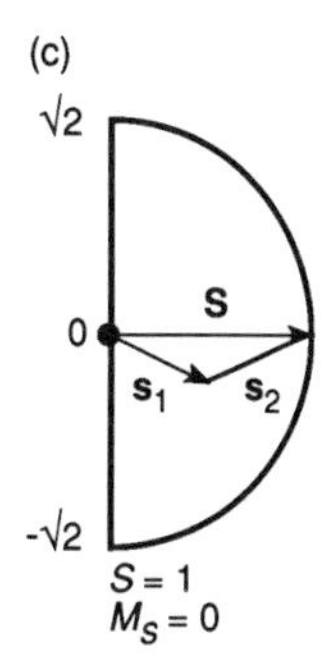

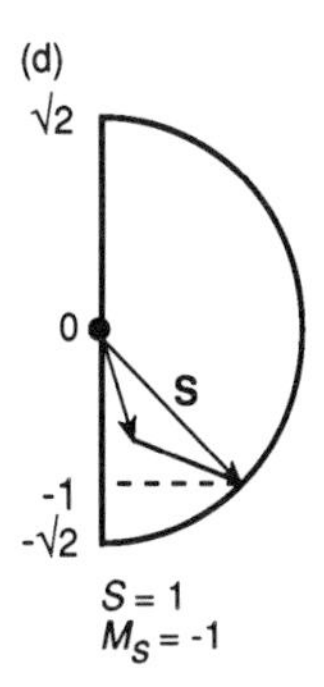

Fig. 4.4 Vector coupling of the two-electron spin angular momenta (units, ħ) to give (a) a singlet state and (b)–(d) a triplet state.

4.5 The Fermi hole

The difference in energy between singlet and triplet states arises from the different probability distributions for the *relative* positions of the two electrons. This is in turn a consequence of the symmetry properties of the *spatial* part of the wavefunction. In the triplet state the spatial wavefunction, Ψ_{space} is antisymmetric with respect to electron exchange and we may write the probability of finding electron 1 at coordinate $\mathbf{r}_1$ and electron 2 at $\mathbf{r}_2$ as $P(\mathbf{r}_1, \mathbf{r}_2)\mathrm{d}\mathbf{r}_1\mathrm{d}\mathbf{r}_2$, where,

$$P(\mathbf{r}_1, \mathbf{r}_2) = \mid \Psi_{\text{space}}(\mathbf{r}_1, \mathbf{r}_2) \mid^2 = \frac{1}{2}[(\phi_{1s}(\mathbf{r}_1)\phi_{2s}(\mathbf{r}_2) - \phi_{1s}(\mathbf{r}_2)\phi_{2s}(\mathbf{r}_1)]^2. \tag{4.35}$$

If $\mathbf{r}_1 = \mathbf{r}_2 = \mathbf{r}$ then

$$P(\mathbf{r}, \mathbf{r}) = \frac{1}{2}[\phi_{1s}(\mathbf{r})\phi_{2s}(\mathbf{r}) - \phi_{1s}(\mathbf{r})\phi_{2s}(\mathbf{r})]^2 = 0. \tag{4.36}$$

The antisymmetric nature of the wavefunction ensures that the probability of finding the two electrons at the same point in space is zero. This minimum in the relative probability distribution for the two electrons in the triplet state is known as a *Fermi hole*. For the singlet wavefunction the spatial wavefunction is symmetric and the probability function does not cancel out at $\mathbf{r}_1 = \mathbf{r}_2$.

Figure 4.5 shows a sketch of the probability distribution of electron 2 assuming electron 1 is fixed at an arbitrary position $\mathbf{r}_1$. The Fermi hole for the triplet wavefunction is clearly shown, and the singlet wavefunction shows a Fermi heap, a maximum in the probability distribution at $\mathbf{r}_1 = \mathbf{r}_2$.

4.6 The exchange and coulomb integrals

The electron–electron repulsion energy can be expressed in terms of the so-called *coulomb* and *exchange* integrals, showing mathematically the origin of the triplet–singlet energy splitting. The total Hamiltonian operator for the helium atom may be written in the form

$$\mathcal{H} = \mathcal{H}_1^0 + \mathcal{H}_2^0 + \mathcal{H}_{12}, \tag{4.37}$$

where $\mathcal{H}_i^0$ has the same form as the one-electron Hamiltonian for a hydrogen-like atom,

$$\mathcal{H}_i^0 = -\frac{\hbar^2}{2\mu^2}\nabla_i^2 - \frac{Ze^2}{4\pi\epsilon_0 r_i}, \tag{4.38}$$

and the electron–electron repulsion operator has the form

$$\mathcal{H}_{12} = +\frac{e^2}{4\pi\epsilon_0 r_{12}} \qquad r_{12} = \mid \mathbf{r}_1 - \mathbf{r}_2 \mid . \tag{4.39}$$

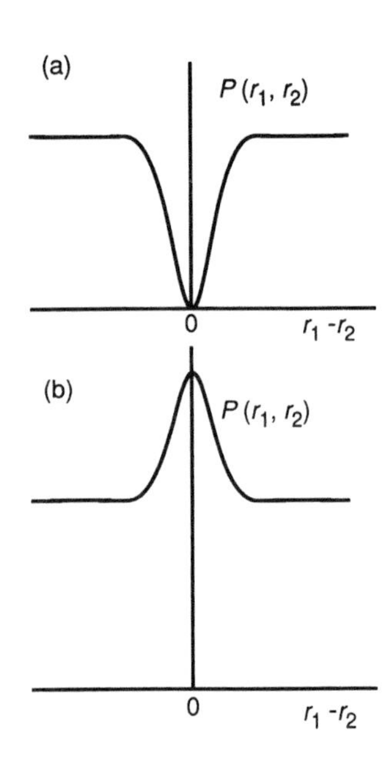

Fig. 4.5 Schematic probability distribution for electron 2 relative to the position of electron 1. (a) For a triplet state (antisymmetric spatial wavefunction) a Fermi hole is observed. (b) For a singlet state (symmetric spatial wavefunction) there is a Fermi heap.

It is then assumed that the solutions, Ψ, of the Schrödinger equation $\mathcal{H}\Psi = E\Psi$ are approximately given by eqns (4.30) with the one-electron orbitals being hydrogen-like eigenfunctions of the one-electron Hamiltonian (4.38). For these approximate wavefunctions, the expectation value (see Section 1.4, eqn 1.18) of the Hamiltonian operator $\langle\mathcal{H}\rangle$ gives an estimate of the energy,

$$\begin{aligned} E = \langle\mathcal{H}\rangle &= \langle\mathcal{H}_1^0 + \mathcal{H}_2^0 + \mathcal{H}_{12}\rangle & (4.40)\\ &= \langle\mathcal{H}_1^0\rangle + \langle\mathcal{H}_2^0\rangle + \langle\mathcal{H}_{12}\rangle & (4.41)\\ &= E_1 + E_2 + \langle\mathcal{H}_{12}\rangle. & (4.42) \end{aligned}$$

E_1 and E_2 are just the one-electron orbital energies—for the $1s2s$ configuration, these are the hydrogenic orbital energies for He^+ in the $1s$ and $2s$ states. The electron–electron repulsion energy $\langle\mathcal{H}_{12}\rangle$ can be expressed in terms of two parameters as follows. Ignoring the constant factor $e^2/4\pi\epsilon_0$, the electron–electron repulsion energy is just the expectation value, $\langle r_{12}^{-1}\rangle$. Using the wavefunctions given above in (4.30), we obtain;

$$\begin{aligned} \langle r_{12}^{-1}\rangle &= \frac{1}{2}\int (\phi_{1s}(\mathbf{r}_1)\phi_{2s}(\mathbf{r}_2) \pm \phi_{1s}(\mathbf{r}_2)\phi_{2s}(\mathbf{r}_1) \times \\ &\quad \times \frac{1}{r_{12}}(\phi_{1s}(\mathbf{r}_1)\phi_{2s}(\mathbf{r}_2) \pm \phi_{1s}(\mathbf{r}_2)\phi_{2s}(\mathbf{r}_1)\mathrm{d}\tau \\ &= J \pm K, \end{aligned} \tag{4.43}$$

where

$$J = \int \phi_{1s}(\mathbf{r}_1)\phi_{2s}(\mathbf{r}_2)\frac{1}{r_{12}}\phi_{1s}(\mathbf{r}_1)\phi_{2s}(\mathbf{r}_2)\mathrm{d}\tau \quad (4.44)$$

$$= \int \phi_{1s}(\mathbf{r}_2)\phi_{2s}(\mathbf{r}_1)\frac{1}{r_{12}}\phi_{1s}(\mathbf{r}_2)\phi_{2s}(\mathbf{r}_1)\mathrm{d}\tau, \quad (4.45)$$

and is called the *coulomb* integral. It is the electron–electron repulsion energy that would be expected if antisymmetrization of the total wavefunction was ignored, i.e., if the wavefunction was just $\phi_{1s}(\mathbf{r}_1)\phi_{2s}(\mathbf{r}_2)$.

$$K = \int \phi_{1s}(\mathbf{r}_1)\phi_{2s}(\mathbf{r}_2)\frac{1}{r_{12}}\phi_{1s}(\mathbf{r}_2)\phi_{2s}(\mathbf{r}_1)\mathrm{d}\tau \quad (4.46)$$

$$= \int \phi_{1s}(\mathbf{r}_2)\phi_{2s}(\mathbf{r}_1)\frac{1}{r_{12}}\phi_{1s}(\mathbf{r}_1)\phi_{2s}(\mathbf{r}_2)\mathrm{d}\tau. \quad (4.47)$$

K is known as the *exchange* integral. As both these quantities, J and K, are always positive and the + and − signs are applicable to the singlet and triplet state respectively, the singlet has higher energy than the triplet state, the energy splitting being equal to $2Ke^2/4\pi\epsilon_0$.

4.7 Configuration interaction and double excitations

To this point in our discussion of atoms with more than one electron, we have assumed the validity of the orbital approximation; a specified number of electrons is assigned to each orbital, and the total wavefunction is expressed as a product of one-electron orbitals, subject to the electron exchange requirements of the Pauli principle. However, attempts to solve the Schrödinger equation using the self-consistent field approach do not always accurately reproduce the experimentally determined energy levels, even when interactions such as the spin–orbit coupling are taken into account. Physically, these deviations are primarily associated with the failure to correctly treat *electron correlation*. The orbital approximation does not take full account of the dependence of the electron–electron interaction on the *relative* instantaneous positions of the electrons. The reality of electron motion is in accord with the tendency to minimize the energy; the energy of the atom would be lower if the electronic motion was arranged so that electrons took paths that instantaneously kept themselves apart as much as possible.

We have already seen that the Pauli principle indroduces the idea of spin correlation, which for the triplet states prevents the electrons from coming very close together. However, to accurately reproduce energy levels, more correlation must be built into both the singlet and triplet state wavefunctions, by acknowledging the breakdown of the orbital approximation. In quantum mechanical calculations the breakdown is incorporated by a purely mathematical operation known as configuration interaction (CI), in which it is assumed that the electron configurations obtained within the orbital approximation are a suitable basis for expanding the wavefunction in the form of a linear combination.

So for example, the ground state of the helium atom is more correctly described by a wavefunction written as

$$\Psi = a\,\Psi(1s^2) + b\,\Psi(2s^2) + c\,\Psi(1s2s) + \ldots\,, \tag{4.48}$$

where $\Psi(2s^2)$, etc., implies an antisymmetrized wavefunction of the type described within the orbital approximation, and a, b, c are coefficients. Undoubtedly, for the ground state the coefficient a will be very much larger than any other coefficient, but the other contributions correct the wavefunction for electron correlation effects. In principle, provided we derive the complete set of coefficients for every possible contributing configuration then we will have obtained the exact non-relativistic wavefunction.

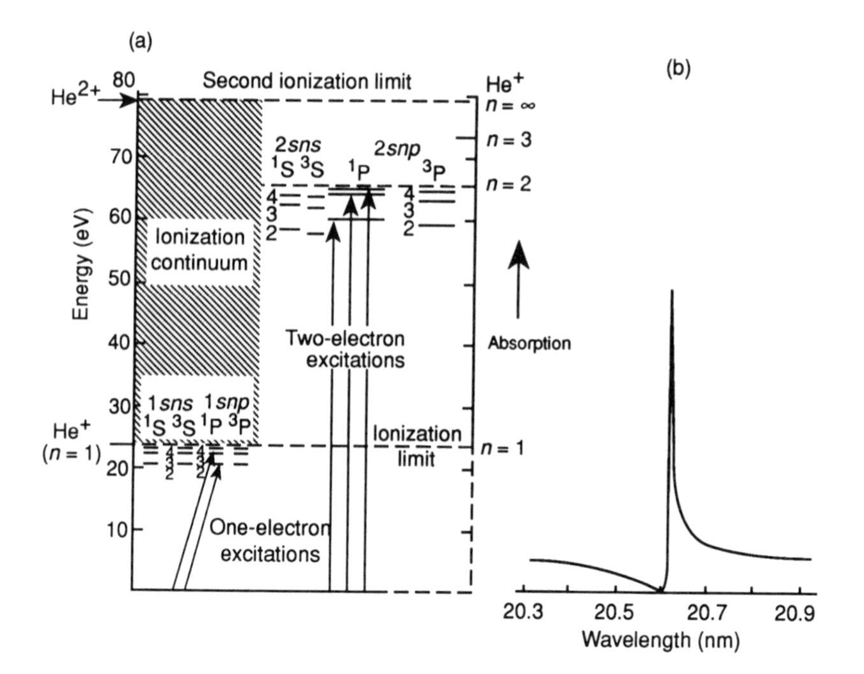

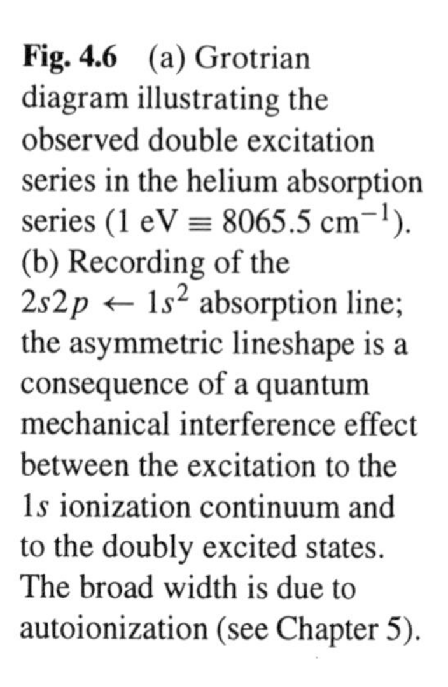
Fig. 4.6 (a) Grotrian diagram illustrating the observed double excitation series in the helium absorption series (1 eV $\equiv$ 8065.5 cm^{-1}). (b) Recording of the $2s2p \leftarrow 1s^2$ absorption line; the asymmetric lineshape is a consequence of a quantum mechanical interference effect between the excitation to the $1s$ ionization continuum and to the doubly excited states. The broad width is due to autoionization (see Chapter 5).

A directly observable consequence of configuration interaction/electron correlation is the appearance of 'double excitations' in the spectrum of the helium atom. Figure 4.6 illustrates some observed transitions which nominally involve the simultaneous excitation of two electrons, e.g.,

$$2s2p \leftarrow 1s^2. \tag{4.49}$$

Such processes are not allowed within the orbital approximation, because one photon can only interact with one electron, hence only one electron can be excited. They can be explained in the CI model, through the small contribution of the $2s^2$ configuration to the nominal $1s^2$ ground state. The transition moment (see Section 2.8) is then given by

$$R_{if} = a \int \Psi^*(1s^2)\hat{\mu}\Psi(2s2p) + b \int \Psi^*(2s^2)\hat{\mu}\Psi(2s2p) + \ldots \tag{4.50}$$

in which the first term is zero but the second term is nonzero. Effectively the transition takes place as a one-electron excitation out of the $2s^2$ part of the ground state wavefunction, $2s2p \leftarrow 2s^2$.

4.8 Lasers II—population inversion in the He–Ne laser

In Section 1.3 the properties of laser radiation were discussed, and it was stated that the observation of laser action required an excess of population in the upper level of a transition, a so-called *population inversion*. Quantum mechanically, the rate of absorption W_{if} calculated for a single molecule in the lower state i, when light is incident at the transition frequency ν with intensity $I(\nu)$ is given by

$$W_{if} = B_{if}\, I(\nu)c, \tag{4.51}$$

where c is the velocity of light and B_{if} is the Einstein coefficient of absorption

$$B_{if} = \frac{1}{6\epsilon_0\hbar^2} \mid R_{if} \mid^2 \qquad R_{if} = \int \Psi_i^* \hat{\mu} \Psi_f \mathrm{d}\tau. \tag{4.52}$$

The rate of stimulated emission for a molecule in its excited state f is exactly the same as expression (4.51) except that the i and f subscripts are reversed. Noting that $R_{if} = R_{fi}$, if we have a sample with N_f molecules in the excited state and N_i in the lower state then the net transition rate is

$$\frac{-\mathrm{d}N_f}{\mathrm{d}t} = (N_f - N_i)B_{if}I(\nu)c. \tag{4.53}$$

In this expression we have, for simplicity, ignored the rate of *spontaneous emission*, a process which tends to remove upper state population without creating laser action. Under conditions of high light intensity, such as are present inside a laser, the spontaneous emission is relatively unimportant. To obtain net stimulated emission, and hence a gain of intensity as light of frequency ν passes through the medium, a population inversion

$$N_f > N_i, \tag{4.54}$$

is required. This is an abnormal state for an atom, for which the equilibrium Boltzmann distribution would under most circumstances give $N_f < N_i$. We therefore need to create a nonequilibrium distribution of population of the electronic states of the atom, and this requires an input of energy, a process known as *pumping*. Unfortunately most methods for pumping are equally efficient at de-exciting the atoms (as is obviously the case for light excitation) and therefore it is impossible to remove more than 50 % of the ground state population into an excited electronic level.

To circumvent this problem, most viable lasers are based on the 'four-level system'. In this set-up, illustrated schematically in Fig. 4.7, the aim is to achieve a population inversion between two levels, neither of which is the ground state, i.e., levels 1 and 2 in the figure. The immediate advantage of this scheme is that, provided the population of level 1 is kept at its normal low level (as

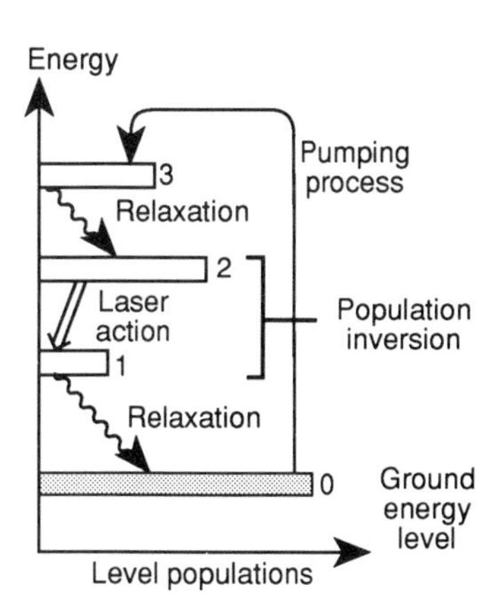

Fig. 4.7 Schematic energy level diagram for a four-level laser system; each level is represented by a box whose length gives the typical relative population.

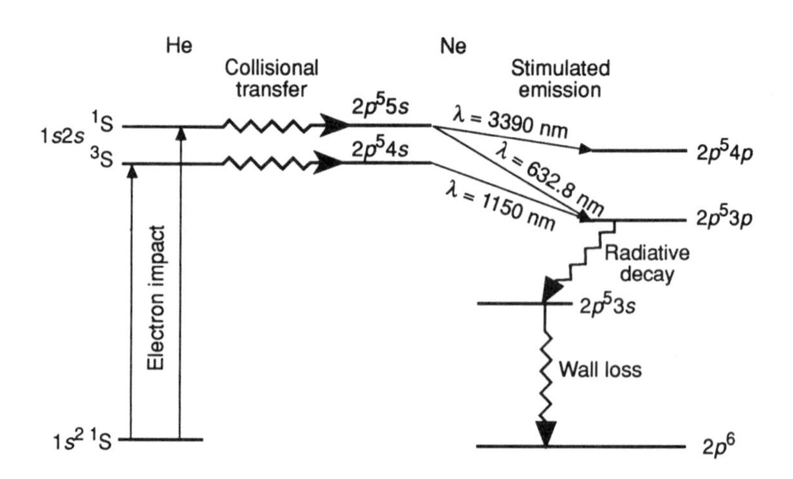

Fig. 4.8 Mechanism for producing a population inversion in the helium–neon laser; splittings of the neon electronic configurations, due to spin–orbit and electrostatic interactions, into terms and levels are not shown.

determined by the Boltzmann distribution), then it is not necessary to move a large percentage of the ground state population into level 2 to achieve an inversion. In the scheme illustrated, a pumping process takes a fraction of the ground state atoms into level 3. Many of the atoms will be pumped back to the ground state 0, but a significant fraction transfer via a *relaxation mechanism* into level 2. It is important that the relaxation does not also populate level 1. Stimulated emission occurs once a population inversion has built up between levels 2 and 1 and then this must be followed by efficient relaxation of atoms in level 1 to the ground state, to remove the population of the lower state involved in the population inversion.

The helium–neon laser illustrates these principles in practice and the appropriate energy level diagram is shown in Fig. 4.8. The laser consists physically of a discharge tube containing a mixture of helium and neon placed between two mirrors to make the laser cavity. When the discharge is struck, energetic helium atoms in the metastable states $1s2s\,^1\mathrm{S}_0$ and $^3\mathrm{S}_1$ are formed by electron impact; the states are long-lived because they cannot fluoresce back to the ground state (for the singlet state this would break the selection rule $\Delta l = \pm 1$, while for the triplet state the $\Delta S = 0$ rule would also be broken). This excitation is equivalent to the pumping process $0 \to 3$ in Fig. 4.7. The helium atoms may undergo subsequent collisions with the neon atoms, and there is a high probability of energy transfer from the helium atom to the neon.

$$\mathrm{He}(1s2s\,^1\mathrm{S}) + \mathrm{Ne}(\ldots 2p^6) \to \mathrm{He}(1s^2) + \mathrm{Ne}\ldots 2p^5\,5s.$$

Such collisional energy transfer processes are very efficient if the relevant energy levels of the acceptor and donor atoms are almost isoenergetic. This is the case for the neon $5s$ state and the metastable singlet state of helium, and likewise for the $4s$ states of neon and the triplet metastable state of helium. The $4s$ and $5s$ states of neon thus become efficiently populated in this relaxation process (equivalent to $3 \to 2$ in Fig. 4.7) while other levels do not. A population inversion is thus created between the $5s$ level of neon and the unpopulated $4p$ and $3p$ levels, and also between the $4s$ and $3p$ levels. Stimulated emission can therefore take place on the three transitions indicated. Finally relaxation occurs in a stepwise fashion, first to the $3s$, then to the $2p$ level via radiative and collisional deactivation processes.

4.9 Problems

1. The wavefunctions for a particle on a circular ring with potential energy $V = 0$ on the ring and $V = \infty$ elsewhere are of the form,

$$\Psi_m = \frac{1}{\sqrt{2\pi}} \exp(im\phi),$$

where m takes integer values $(0, \pm 1, \pm 2 \ldots)$. By analogy with the helium atom construct three space–spin wavefunctions for two electrons confined in the same circular ring, which will obey the Pauli principle, based on the following three spatial wavefunctions:

$$\Psi_{1,1} = \Psi_1(1)\Psi_1(2) \quad (= \frac{1}{2\pi} \exp(i\phi_1) \exp(i\phi_2))$$

$$\Psi^{\pm}_{1,-1} = \frac{1}{\sqrt{2}}[\Psi_1(1)\Psi_{-1}(2) \pm \Psi_{-1}(1)\Psi_1(2)].$$

For each wavefunction determine the probability density $P(0, \phi_2)$ for electron 2 to be at ϕ_2, given that electron 1 is at $\phi_1 = 0$. Comment on the relative energies of the three wavefunctions.

2. A number of possible transitions for the beryllium atom are listed below. Which of these transitions are 'fully allowed'? For the other transitions state which selection rules would have to be broken, and comment on whether there is any mechanism that could cause this breakdown.

$$2s5s(^1S_0) \rightarrow 2s5d(^1D_2)$$
$$2s5s(^3S_1) \rightarrow 2s2p(^1P_1)$$
$$2s5s(^1S_0) \rightarrow 2s^2(^1S_0)$$
$$2s5p(^3P_1) \rightarrow 2s3s(^3S_1)$$
$$2s5p(^3P_1) \rightarrow 3s4s(^3S_1)$$
$$2s3p(^3P_0) \rightarrow 3p4p(^3D_2)$$
$$2s3p(^3P_0) \rightarrow 3p4p(^3P_0)$$

5 The spectra of many-electron atoms

5.1 Coupling of orbital angular momentum

In this chapter we shall consider the spectra of atoms in which there may be more than one electron with orbital angular momentum. An example is the carbon atom, which has the ground state configuration $1s^2 2s^2 2p^2$ and lowest excited states of the type $1s^2 2s^2 2p\,nl$. The total orbital angular momentum vector $\mathbf{L}$ for two unpaired electrons was defined in Section 4.3 as the vector sum of the individual orbital angular momenta, $\mathbf{L} = \mathbf{l}_1 + \mathbf{l}_2$. Consider the excited carbon atom configuration $2p3d$ (ignoring the closed shell $1s$ and $2s$ electrons whose resultant angular momentum is zero); in this case $l_1 = 1$ and $l_2 = 2$. The total orbital angular momentum, if it can be defined at all, must be quantized such that

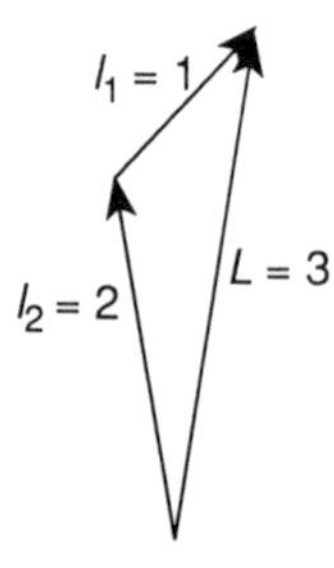

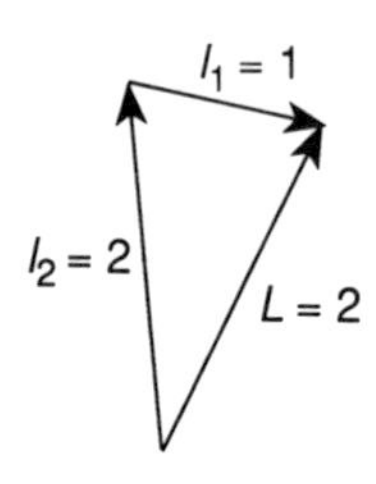

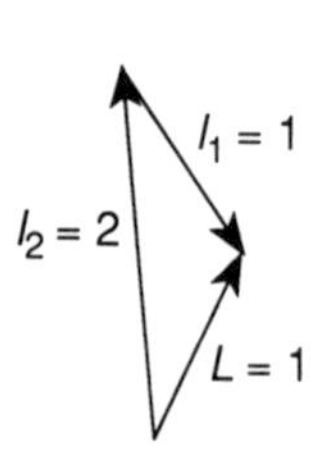

Fig. 5.1 Vector addition of angular momenta with $l_1 = 1$ and $l_2 = 2$. The lengths of the vectors $\mathbf{l}_1, \mathbf{l}_2$ and $\mathbf{L}$ are $\sqrt{l_1(l_1+1)}\hbar$, $\sqrt{l_2(l_2+1)}\hbar$, and $\sqrt{L(L+1)}\hbar$.

$$|\,\mathbf{L}\,| = [L(L+1)]^{\frac{1}{2}}\hbar, \tag{5.1}$$

with the quantum number L integral. Vector addition and subtraction allows a maximum value for L of $l_1 + l_2$ and a minumum value of $|\,l_1 - l_2\,|$. It may also take all intermediate integral values

$$L = l_1 + l_2, l_1 + l_2 - 1, \ldots |\,l_1 - l_2\,|\,. \tag{5.2}$$

For the $2p\,3d$ configuration $L = 3, 2$ or 1 (see Fig. 5.1); the three states are designated F, D and P respectively.

In general, the quantum states of the atom with different L, arising from the same electron configuration, will have different energies, due to the different electrostatic interaction between the electrons. An oversimplified argument can be put forward as follows; a state with $L = l_1 + l_2$ has the orbital angular momenta aligned almost parallel in space (see Fig. 5.1) and effectively, the electrons circulate in the same direction about a common axis. In contrast, the state with $L =|\,l_1 - l_2\,|$ has the angular momenta almost antiparallel, hence the electrons are circulating in opposite directions about a common axis. In the former case the electrons come close to one another rather less often than in the latter, hence there may be greater electron repulsion in the antiparallel alignment.

5.2 Russell–Saunders coupling

In an atom such as carbon, with more than one unpaired electron, there can be up to three types of interaction that lead to states of the same configuration having different energies.

1. Spin correlation; as a consequence of the Pauli principle, discussed in Chapter 4, there is the least electron–electron repulsion for states of the atom in which the unpaired electron spins are aligned to produce the maximum resultant spin angular momentum; for atoms with two unpaired electrons, triplet states generally lie lower in energy than the singlet states arising from the same electron configuration.

2. Coupling of orbital angular momentum; as discussed in Section 5.1 the state of highest L with maximum alignment of orbital angular momenta tends to be lowest in energy (although there are many exceptions, especially when considering excited states).

3. Spin–orbit coupling interaction between the spin and orbital magnetic moments.

An important question is, which of these effects is the largest? There is no unique answer; different atoms may show a different behaviour. The first row atoms of the periodic table, in their low-energy electronic states, follow Russell–Saunders coupling in which the order of importance of the effects is $1 \geq 2 \gg 3$. In energetic terms this may be illustrated as in the schematic diagram, Fig. 5.2a, for an $np\,nd$ configuration. In this case it is assumed $1 \gg 2$, and therefore the singlet and triplet groups of states are well separated in energy. These groups are then split into the *terms* designated using the notation ${}^{2S+1}L$. The terms are split into *levels* by spin–orbit coupling (for the triplet states only). In the Russell–Saunders scheme the resultant orbital angular momentum $\mathbf{L}$ is coupled to the resultant spin $\mathbf{S}$ to give states with total angular momentum vector $\mathbf{J}$ such that the total angular momentum quantum number J takes the values $J = L + S, L + S - 1, \ldots, \mid L - S \mid$. The $np\,nd$ configuration is split into a total of 12 energy levels, each of which is specified by the set of quantum numbers l_1, l_2, L, S, J. The quantum number M_J can also be specified, as discussed below, but the quantum numbers $m_{l_1}, m_{s_1}, m_{l_2}$ and m_{s_2} can no longer be defined; the coupling of the $\mathbf{l}_1$ and $\mathbf{l}_2$ vectors to each other, and also of $\mathbf{s}_1$ and $\mathbf{s}_2$, destroys the space quantization of the individual vectors. Figure 5.2b shows

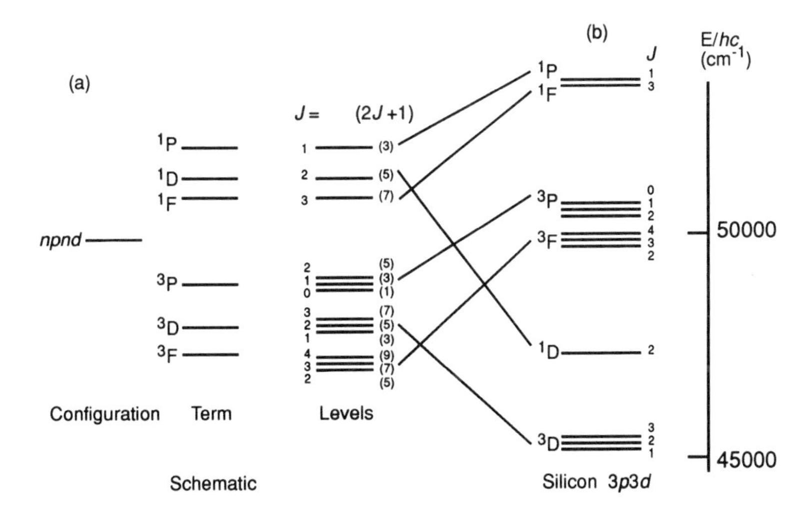

Fig. 5.2 Terms and energy levels in Russell–Saunders coupling. (a) Schematic diagram for $np\,nd$ configuration. The numbers in brackets at the right-hand side give the level degeneracies $(2J + 1)$. (b) Energy levels for the $3p\,3d$ configuration of silicon.

a real example of energy level splittings for the $3p\,3d$ configuration of silicon. In this case the magnitude of effects 1 and 2 are comparable, and there is not a clear splitting into singlet/triplet groups. It is noticeable that in this case the state with the highest L is not lowest in energy, emphasizing the oversimplicity of the arguments presented earlier in Section 5.1, especially when applied to electronically-excited configurations.

Degeneracy

In the Russell–Saunders (RS) coupled representation, each level with total angular momentum quantum number J is permitted $2J + 1$ possible quantized projections of the total angular momentum along the z-axis with corresponding M_J quantum numbers $J, J-1, \ldots, -J$. The numbers in brackets in the last column of Fig. 5.2a give the degeneracy of levels; there are a total of 60 quantum states when the degeneracies are included for the configuration $np\,nd$. This total is also exactly what would be expected if the angular momentum coupling and spin correlation were ignored and we treated the two electrons as independent—the *decoupled representation*. In that case, the quantum numbers $l_1, s_1, m_{l_1}, m_{s_1}$ and $l_2, s_2, m_{l_2}, m_{s_2}$ are well defined, and there are $(2l_1+1)$ possible m_{l_1} states for electron 1, $(2s_1+1)$ m_{s_1} states, etc.; hence the total number of combinations of the magnetic quantum numbers, for $l_1 = 1, s_1 = \frac{1}{2}, l_2 = 2$, and $s_2 = \frac{1}{2}$—the number of *microstates*—is $3 \times 2 \times 5 \times 2 = 60$. This equivalence of the degeneracies between the decoupled and coupled representations must always exist.

Pauli principle restrictions on allowed terms

The derivation of the possible term symbols and levels for configurations where the two unpaired electrons are equivalent, as in the ground state of carbon, $2p^2$, cannot be achieved using only the RS vector coupling arguments given above. If we were to follow that method, we would have $l_1 = l_2 = 1$, implying $L = 2, 1, 0$, and with $S = 1$ or 0 the term symbols should be (ignoring spin–orbit coupling)

$$^3\mathrm{D},\ ^1\mathrm{D},\ ^3\mathrm{P},\ ^1\mathrm{P},\ ^3\mathrm{S},\ ^1\mathrm{S}. \tag{5.3}$$

However, this procedure ignores the Pauli principle and the equivalence of the two electrons. The total number of microstates in the *decoupled* representation does not equal $(2l_1+1)(2l_2+1)(2s_1+1)(2s_2+1)$ because, for example, the state

$$m_{l_1} = 1\ m_{s_1} = \frac{1}{2} \qquad m_{l_2} = 1\ m_{s_2} = \frac{1}{2}, \tag{5.4}$$

is not allowed; both electrons would have the same set of four quantum numbers n, l, m_l and m_s. Also the two possibilities

$$m_{l_1} = 1\ m_{s_1} = \frac{1}{2} \qquad m_{l_2} = -1\ m_{s_2} = -\frac{1}{2} \tag{5.5}$$

$$m_{l_1} = -1\ m_{s_1} = -\frac{1}{2} \qquad m_{l_2} = 1\ m_{s_2} = \frac{1}{2}, \tag{5.6}$$

should only be counted as one, because the two electrons are indistingishable and these do not represent two different microstates of the atom. Note that such arguments would not apply if we had the configuration $2p\,3p$, for then the two electrons are inequivalent. These effects reduce the number of permitted microstates in the decoupled representation to 15 as listed in Table 5.1.

The total degeneracy of the six terms listed in eqn (5.3) is $15 + 5 + 9 + 3 + 3 + 1 = 36$; hence, bearing in mind the need for equivalence between the degeneracies in the RS coupled and decoupled representations, it is clear that not all these terms can exist. To determine which terms are allowed or forbidden we make use of one more criterion, that the number of states in the coupled representation with a particular pair of values for M_S and M_L must equal the number of microstates in the decoupled representation for which $m_{s_1} + m_{s_2} = M_S$ and $m_{l_1} + m_{l_2} = M_L$.

To show how this works in practice, the allowed decoupled states are listed in Table 5.1 in a format that groups them according to the sums $m_{s_1} + m_{s_2}$ and

Table 5.1

Microstates ($m_{l_1}\, m_{s_1}\, m_{l_2}\, m_{s_2}$) in the decoupled representation for an np^2 configuration

	$M_L = m_{l_1} + m_{l_2}$				
$M_S = m_{s_1} + m_{s_2}$	$M_L = 2$	$M_L = 1$	$M_L = 0$	$M_L = -1$	$M_L = -2$
$M_S = 1$		$(1\,\frac{1}{2}\,0\,\frac{1}{2})$	$(1\,\frac{1}{2}\,-1\,\frac{1}{2})$	$(-1\,\frac{1}{2}\,0\,\frac{1}{2})$	
$M_S = 0$	$(1\,\frac{1}{2}\,1\,\frac{-1}{2})$	$(1\,\frac{1}{2}\,0\,\frac{-1}{2})$ $(1\,\frac{-1}{2}\,0\,\frac{1}{2})$	$(1\,\frac{1}{2}\,-1\,\frac{-1}{2})$ $(1\,\frac{-1}{2}\,-1\,\frac{1}{2})$ $(0\,\frac{-1}{2}\,0\,\frac{1}{2})$	$(-1\,\frac{1}{2}\,0\,\frac{-1}{2})$ $(-1\,\frac{-1}{2}\,0\,\frac{1}{2})$	$(-1\,\frac{1}{2}\,-1\,\frac{-1}{2})$
$M_S = -1$		$(1\,\frac{-1}{2}\,0\,\frac{-1}{2})$	$(1\,\frac{-1}{2}\,-1\,\frac{-1}{2})$	$(-1\,\frac{-1}{2}\,0\,\frac{-1}{2})$	

$m_{l_1} + m_{l_2}$. (Note that the use of M_S and M_L in this method implies that the spin–orbit coupling is ignored: otherwise, we would have to group the microstates according to the value of M_J.) A ^{3}D state would have a degeneracy of 15 which would be all the combinations of $M_L = +2, 1, 0, -1, -2$ and $M_S = 1, 0, -1$. However, in the decoupled representation there is no component permitted under the Pauli principle with $m_{l_1} + m_{l_2} = 2$ *and* $m_{s_1} + m_{s_2} = 1$. Therefore the ^{3}D term is not permitted. There must be at least one state with $M_L = 2$, $M_S = 0$ and therefore this must arise from the ^{1}D term, degeneracy 5. We now delete one microstate from each column of the middle row of the table to account for the five M_L components, $M_S = 0$, $M_L = \pm 2, \pm 1, 0$ leaving ten microstates unaccounted for. The ^{3}P state has nine components with $M_L = +1, 0, -1$, $M_S = +1, 0, -1$, and this is the only term of the four remaining possibilities in eqn (5.3) to have $M_L = \pm 1$, $M_S = \pm 1$ components. These must correspond to the microstates in the appropriate columns of the table and therefore this term must also exist. We then remove another nine microstates from the table, one from each row of the middle three columns, leaving just one remaining microstate with $M_S = 0$, $M_L = 0$. Therefore the remaining permitted term must be ^{1}S.

In summary, from the configuration $2p^2$, we have shown that the only permitted terms are ^{1}D, ^{3}P, and ^{1}S, degeneracy $5 + 9 + 1 = 15$. Similar arguments could be used to show that the terms arising from a d^2 configuration would be ^{1}G, ^{3}F, ^{1}D, ^{3}P and ^{1}S.

Hund's rules

Hund's rules are used to predict the lowest energy term and level arising from the ground electron configuration in cases where Russell–Saunders coupling applies. The rules are based on the relative magnitudes of the various interactions listed at the beginning of this section and should be applied in the following order of priority, i.e., $1 > 2 > 3$.

Table 5.2
Term symbols for first-row atoms

Element	Configuration	Terms arising	Ground state level
Li	$1s^2\,2s$	^{2}S	$^2S_{\frac{1}{2}}$
Be	$1s^2\,2s^2$	^{1}S	$^1S_0$
B	$1s^2\,2s^2\,2p$	^{2}P	$^2P_{\frac{1}{2}}$
C	$1s^2\,2s^2\,2p^2$	^{1}D, ^{3}P, ^{1}S	$^3P_0$
N	$1s^2\,2s^2\,2p^3$	^{4}S, ^{2}D, ^{2}P	$^4S_{\frac{3}{2}}$
O	$1s^2\,2s^2\,2p^4$	^{1}D, ^{3}P, ^{1}S	$^3P_2$
F	$1s^2\,2s^2\,2p^5$	^{2}P	$^2P_{\frac{3}{2}}$
Ne	$1s^2\,2s^2\,2p^6$	^{1}S	$^1S_0$

1. The state with the highest spin multiplicity will be lowest in energy.
2. If there is more than one term with the highest spin multiplicity, the term with the highest L will be lowest in energy.
3. For a term giving rise to more than one level, the lowest J value is lowest in energy if the outermost subshell is less than half full, but the highest J value is lowest for more than half-full subshells (e.g., p^4, p^5, d^6, d^7 ...).

Applying these rules for the carbon atom, ground configuration $2p^2$, the ^{3}P term has the highest spin multiplicity and is therefore lower in energy than ^{1}D or ^{1}S. The subshell is less than half full so the $^3P_0$ state is the lowest energy level. For oxygen, ground configuration $2p^4$ the terms ^{1}D, ^{3}P, ^{1}S can also be obtained making use of a general rule that the configuration $2p^{6-n}$ is equivalent to $2p^n$ in respect of the term symbols arising from it. However, the $^3P_2$ level is now lowest in energy, applying Hund's third rule. In Table 5.2 the ground state electron configurations, the possible terms arising, and the experimentally-determined lowest energy level are given for the first row atoms, and it can be seen that Hund's rules are successful in these cases. It must be noted that Hund's rules are only strictly applicable to finding the *lowest* energy level of atoms obeying Russell–Saunders coupling. In many cases the observed energy ordering for excited electronic configurations is different from that predicted by Hund's rules.

Figure 5.3 illustrates the various transitions that are observed in the spectrum of the carbon atom from the lowest energy configuration $2s^2\,2p^2$. The transitions with bold lines obey the selection rules, $\Delta l = \pm 1$, $\Delta S = 0$, $\Delta J = 0, \pm 1$, $\Delta L = 0, \pm 1$, $\Delta M_J = 0, \pm 1$. Note that a transition obeying $\Delta L = 0$ is permissible, provided that the one-electron selection rule, $\Delta l = \pm 1$ is obeyed, e.g., $2s^2\,2p^2\,(^3\mathrm{P}) \rightarrow 2s^2\,2p\,ns\,(^3\mathrm{P})$. Some *intercombination* lines involving $\Delta S \neq 0$ have also been observed very weakly in the emission spectrum, for

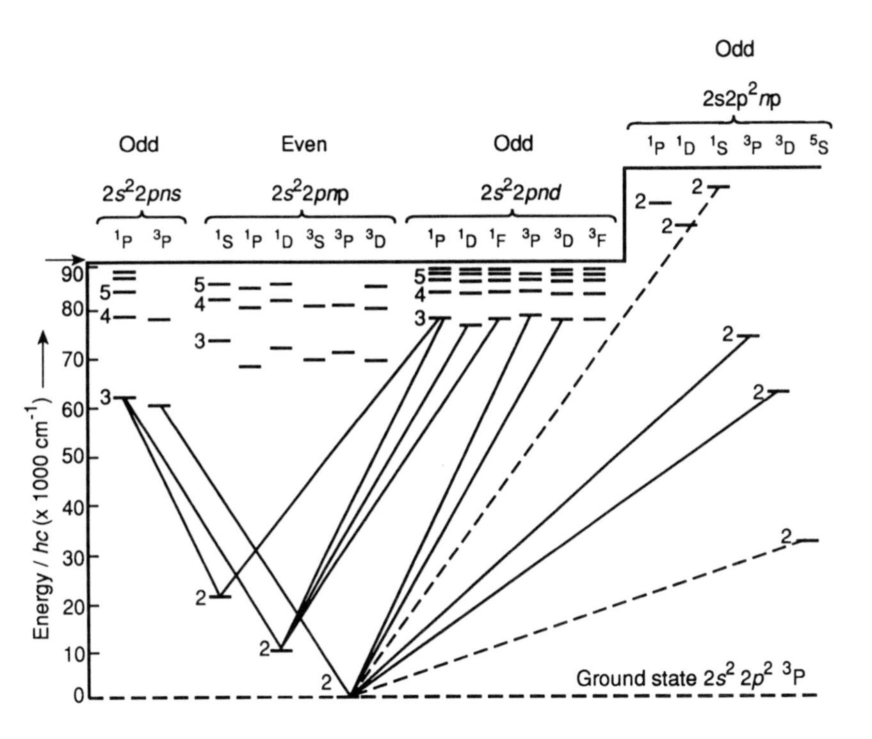

Fig. 5.3 Grotrian diagram of carbon. From the ground state configuration, only transitions to excited configurations with odd symmetry overall are allowed. Each line actually represents a series, e.g., $2s^2 2p\, ns\,^3\mathrm{P}$), $n = 3, 4, 5 \ldots \leftarrow 2s^2 2p^2 (^3\mathrm{P})$

example $^5\mathrm{S} \rightarrow {}^3\mathrm{P}$ (indicated with dashed lines). The observation of the intercombination lines is evidence for a minor breakdown of the Russell–Saunders coupling scheme, as is discussed in the next section.

5.3 jj-coupling

The Russell–Saunders coupling scheme is valid provided that the spin–orbit interaction is much smaller than the other effects due to electron–electron repulsion. The magnitude of spin–orbit interaction increases with the nuclear charge and therefore the validity of this scheme becomes more doubtful for atoms in the lower rows of the periodic table. It may also become invalid, even for first-row atoms in their highly excited states. When we say the coupling scheme is 'valid' we mean that it correctly identifies the 'good' quantum numbers in the atom—l_1, l_2, L, S, J, M_J in the Russell–Saunders scheme—and that the total wavefunction is an eigenfunction of certain angular momentum operators, $\hat{\mathbf{l}}_1^2, \hat{\mathbf{l}}_2^2, \hat{\mathbf{L}}^2, \hat{\mathbf{S}}^2, \hat{\mathbf{J}}^2$, and $\hat{J}_z$. In reality the wavefunction may not be an exact eigenfunction of these operators but RS coupling would still be a good starting point provided the true wavefunction was very similar to an exact eigenfunction. Mathematically this could be described by saying that the true wavefunction was a linear combination of the full set of RS coupled states, but with a very dominant contribution from one state in particular.

For some of the heavier elements, however, the spin–orbit interaction is so large that RS coupling is not even approximately valid, and a better starting point for describing the states is to use *jj-coupling*. In this scheme each unpaired electron spin is coupled directly by spin–orbit coupling to its own orbital angular momentum to produce a local total angular momentum j. For example, if the unpaired electrons are described by the configuration $5p\,6s$ then we have

$$s_1 = \frac{1}{2}, l_1 = 1 \quad \Rightarrow \quad j_1 = \frac{1}{2} \text{ or } \frac{3}{2} \tag{5.7}$$

$$s_2 = \frac{1}{2}, l_2 = 0 \quad \Rightarrow \quad j_2 = \frac{1}{2} \text{ only.} \tag{5.8}$$

The angular momenta of the individual electrons are coupled as though there were no other unpaired electrons present. In some cases electrostatic effects may lead to a subsequent coupling of the two individual total angular momenta

$$\mathbf{J} = \mathbf{j}_1 + \mathbf{j}_2, \tag{5.9}$$

and an overall total angular momentum quantum number J is thus defined, taking values $j_1+j_2, j_1+j_2-1, \ldots |j_1-j_2|$. The primary splittings of the energy levels are associated with the spin–orbit coupling for each electron, and then there are further small splittings due to electrostatic effects and spin correlation. Note, however, that the quantum numbers L and S are not defined, the 'good' quantum numbers being $l_1, s_1, j_1, l_2, s_2, j_2, J, M_J$. Figure 5.4 shows the difference in energy level patterns for the configurations np^2 and $np\,(n+1)s$ in

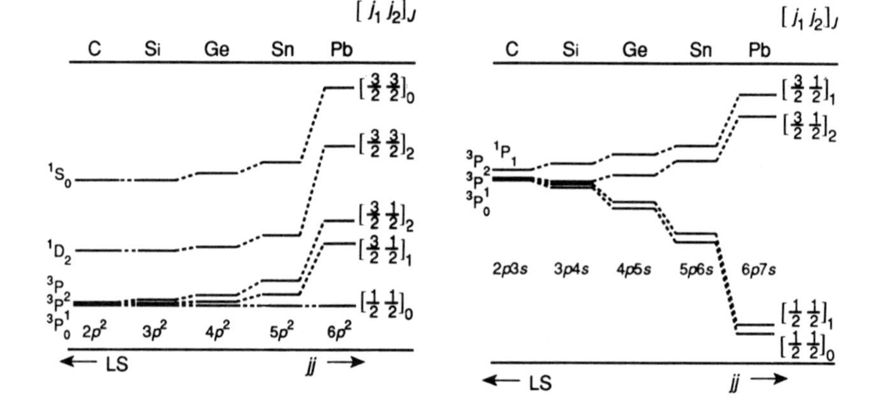

Fig. 5.4 Transition from LS- to jj-coupling as shown by the ground states (np^2), and first excited states ($np\ (n+1)s$) of the carbon group of elements.

the carbon group of the elements. For carbon, obeying Russell–Saunders coupling, the energies are primarily determined by L and S, whereas in lead, they are determined by the two values of j_1 and j_2.

Once again it is unlikely that any atom will exactly obey jj-coupling, but this scheme may be a close representation of the true wavefunction; i.e., an expansion of the true wavefunction in terms of the full set of jj-coupled states will give one dominant contribution to the expansion. It should be noted that any exact Russell–Saunders state can be expressed as a linear combination of a full set of jj-coupled states arising from the same configuration, and vice versa. Therefore jj-coupling and RS coupling can simply be regarded as two limiting cases of a range of different relative magnitudes for the electrostatic and spin–orbit interactions; in any specific case the states are described by the notation for the coupling scheme which is closest to the true situation. For example, the np^2 configurations of germanium and tin clearly show an intermediate behaviour as illustrated in Fig. 5.4.

In the spectra of atoms obeying jj-coupling, the selection rules applying are

$$\Delta j_1 = 0 \quad \Delta j_2 = 0, \pm 1 \quad \text{or} \tag{5.10}$$

$$\Delta j_2 = 0 \quad \Delta j_1 = 0, \pm 1 \quad \text{and} \tag{5.11}$$

$$\Delta J = 0, \pm 1. \tag{5.12}$$

It should be noted that there are no selection rules governing the change of L or S as these quantum numbers are undefined. The appearence of intercombination lines (disobeying $\Delta S = 0$) in the spectra of atoms that are essentially Russell–Saunders coupled, is indicative that the true wavefunctions have some intermediate character between RS and jj-coupling, e.g., for carbon there is a slight admixture of $S = 0$ and $S = 2$ character into the $S = 1$ triplet wavefunctions.

5.4 The Zeeman and Stark effects

In earlier sections the spin–orbit effect has been identified as an important *internal* interaction within an atom, which causes a splitting of energy levels. When an atom is placed in an *external* magnetic or electric field, this is also likely to cause a perturbation to the energy levels of the system. The *Zeeman effect* describes what normally happens in a *magnetic* field whereas the *Stark effect* is concerned with an external *electric* field. In this text we shall only be considering *homogeneous* fields, i.e., those in which all the atoms in the sample experience the same field irrespective of their position in space; the use of *inhomogeneous* fields is very important in other contexts (see the Stern–Gerlach experiment, in Section 2.10, for instance).

The M_J quantum number in an atom defines the projection of the total angular momentum of the atom with reference to a laboratory fixed z-axis. The choice of orientation of this axis is arbitrary until we expose the atom to an external field. The laboratory directions then become inequivalent, and the quantization axis is conveniently defined as the field direction. Considering first the Zeeman effect, the M_J quantum states correspond to different orientations

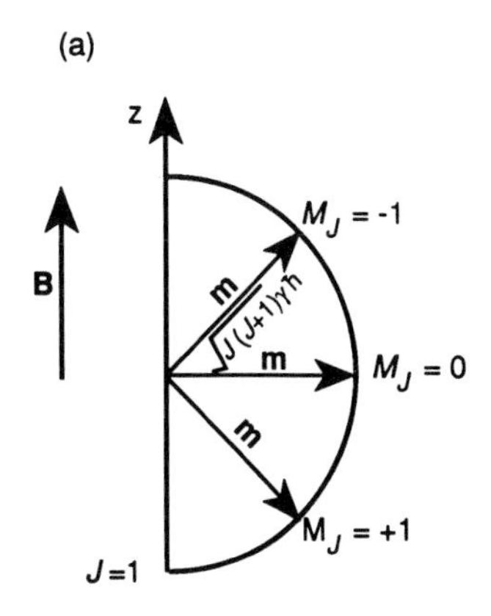

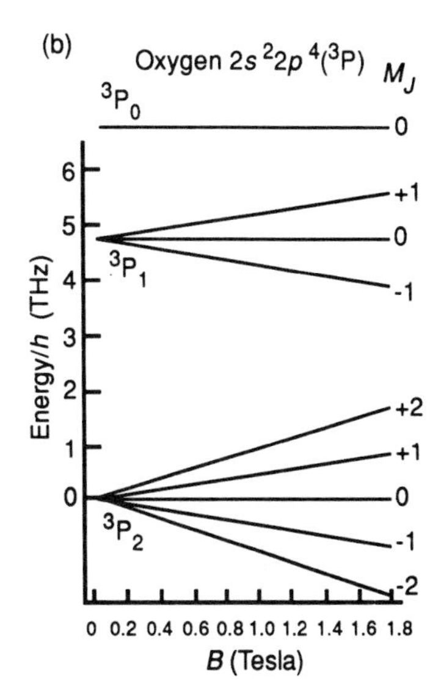

Fig. 5.5 (a) Orientation of the magnetic moment vector **m** for the three M_J states for $J = 1$ relative to the magnetic field direction. **m** points in the opposite direction to **J**. The field is applied along the z-axis. As the x, y components of **m** are not well defined, the vector is considered to precess around the z-direction. (b) Splitting of the ground term of the oxygen atom as a function of magnetic field.

of the magnetic moment of the atom with respect to the applied field. For the state $M_J = +J$, the magnetic moment is almost antiparallel with the external magnetic field direction (Fig. 5.5a), whereas for the state $M_J = -J$, the magnetic moment and field are almost parallel. Classically, the energy of a magnetic dipole **m** in a field **B** is given by $E = -\mathbf{m}.\mathbf{B}$, with the parallel alignment being of lowest energy and the antiparallel alignment of highest energy. Therefore, the various M_J states become *non-degenerate*; the energies of the M_J states for the ^{3}P term of the oxygen atom are shown as a function of magnetic field in Fig. 5.5b.

The Stark effect is somewhat different in its characteristics, because an atom does not possess an *electric* dipole moment, and therefore does not interact directly with an external electric field. Nevertheless, the Stark effect can still lead to the lifting of the M_J degeneracy of a particular level, although in most cases the energy depends on $|M_J|$ rather than M_J itself. The Stark effect occurs indirectly as a result of the field *inducing* a dipole moment, by perturbing the electronic charge distribution. The dipole may then have a favourable electrostatic interaction with the field, lowering its energy. The M_J dependence of the interaction arises from the differences in the orientation of the electronic charge distribution in the different states with respect to the quantization axis; the magnitude of the induced dipole is dependent on this orientation. For example, in the term $ns\,np$ (^{1}P) the $M_J = 0$ component would have the p orbital directed along the z-axis (pure p_z orbital), whereas the $M_J = \pm 1$ components will be a mixture of p_x and p_y orbitals. We will not deal in detail with the Stark effect but note that for relatively small fields, F, the observed energy level splittings are proportional to F^2 (except for the hydrogen atom). The quadratic dependence is illustrated in Fig. 5.6; the $5p\ ^2\mathrm{P}_{\frac{3}{2}}$ state of potassium is split into

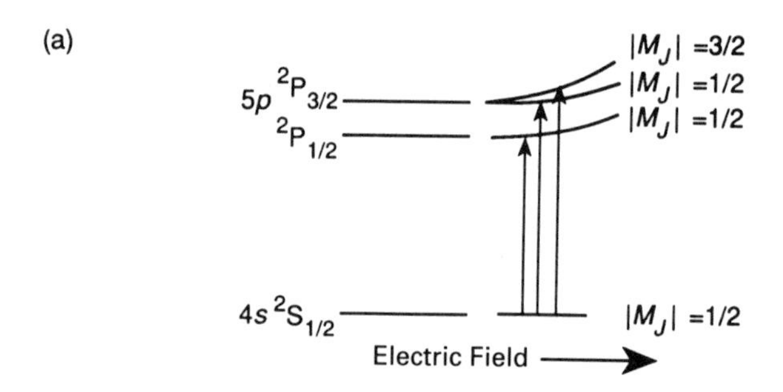

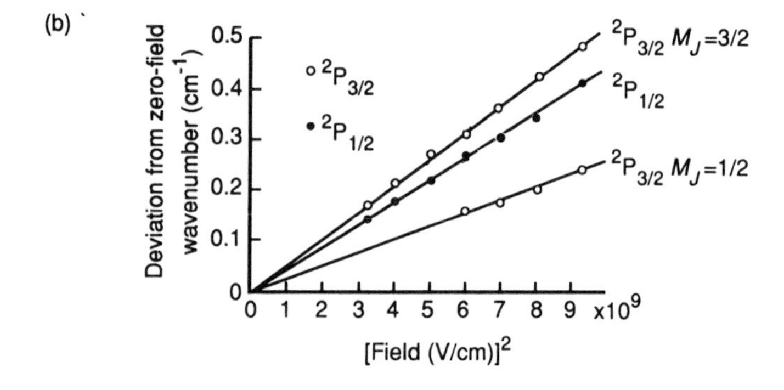

Fig. 5.6 The Stark effect for potassium. (a) Schematic dependence of the $4s$ and $5p$ energy levels on the electric field. (b) A graph plotting the deviation from zero-field positions of the $5p\ ^2\mathrm{P}_{\frac{1}{2},\frac{3}{2}} \leftarrow 4s\ ^2\mathrm{S}_{\frac{1}{2}}$ transition wavenumbers against the square of the electric field.

two components, with $|M_J| = \frac{3}{2}$ and $\frac{1}{2}$, leading to the deviations from zero-field energy that are shown. The $^2P_{\frac{1}{2}}$ component is not split, as $M_J = \pm\frac{1}{2}$ only, but it does undergo an energy shift, which also has a quadratic field dependence. The quadratic behaviour reflects the fact that the field performs two tasks; first to induce the dipole and secondly to interact with that dipole. It should be noted that atoms in states with no resultant total angular momentum, i.e., $J = 0$, have no Stark or Zeeman effects to a first-order approximation.

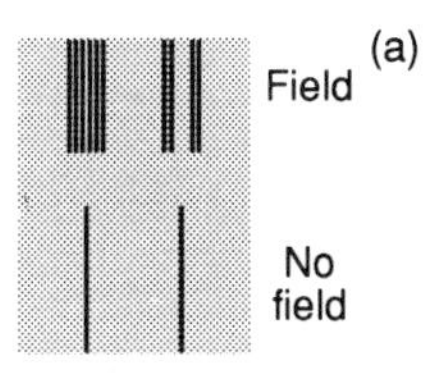

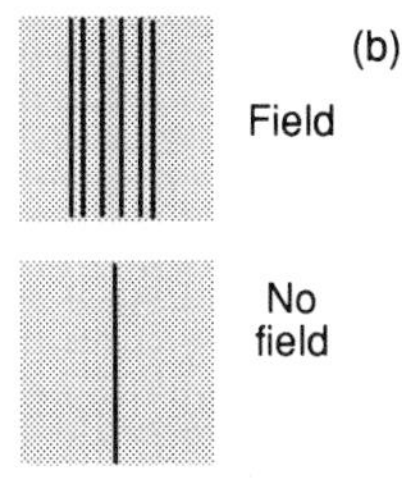

Fig. 5.7 The Zeeman effect for (a) sodium $3p\,^2P_{\frac{1}{2},\frac{3}{2}} \rightarrow 3s\,^2S_{\frac{1}{2}}$, and (b) zinc $^3P_1 \rightarrow\, ^3S_1$.

Selection rules

The perturbation of energy levels by an external field leads to a splitting of lines in the atomic spectra. Figure 5.7a shows the sodium 'D-lines' ($3p\,^2P \rightarrow 3s\,^2S$) observed in zero field and in the presence of a magnetic field, while Fig. 5.7b makes the same comparison for the $^3P_1 \rightarrow\, ^3S_1$ transition of zinc; in the latter case a single line at zero field becomes six lines with the magnetic field applied. To understand the appearance of the spectra, the selection rules must be reconsidered. The selection rule

$$\Delta M_J = 0, \pm 1, \tag{5.13}$$

appeared to be relatively insignificant when it was introduced in Section 5.2. However, it becomes very important in the observation of the Stark or Zeeman effects. For a normal electric-dipole allowed transition, the $\Delta M_J = 0$ selection rule only applies to the specific case when the polarization of the exciting light is *parallel* to the electric or magnetic field, whereas the $\Delta M_J = \pm 1$ rule applies to *perpendicular* orientation. This subtlety is important when a polarized, well-directed light source, such as a laser is employed. The transitions that give rise to the spectra of Fig. 5.7 are illustrated in Fig. 5.8.

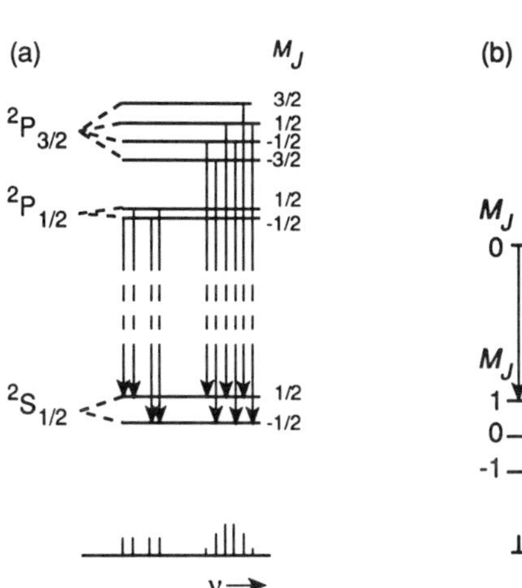

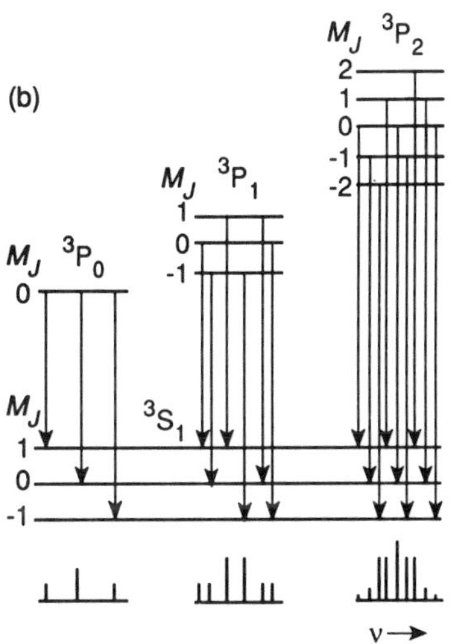

Fig. 5.8 Energy level diagram illustrating the observed transitions in the Zeeman effect for (a) sodium $3p\,^2P_{\frac{1}{2},\frac{3}{2}} \rightarrow 3s\,^2S_{\frac{1}{2}}$ and (b) zinc $^3P \rightarrow\, ^3S$.

5.5 Quantitative description of the Zeeman effect

The normal Zeeman effect

The *normal Zeeman effect* occurs for atoms that have no resultant spin angular momentum, but have a total resultant orbital angular momentum **L**. The magnetic moment of the atom is given by

$$\mathbf{m} = \gamma_e \mathbf{L} \qquad \gamma_e = \frac{-e}{2m_e}, \tag{5.14}$$

and the classical energy of interaction with the magnetic field is

$$E = -\mathbf{m}.\mathbf{B} = -\gamma_e \mathbf{L}.\mathbf{B}. \tag{5.15}$$

In quantum mechanics, we can replace the quantities in eqn (5.15) with the corresponding operators, and then determine the expectation value of the operator product to obtain the Zeeman energy (in a first-order approximation, assuming the wavefunction is not perturbed by the magnetic field). This energy is then added to the field-free energy level E_0,

$$E = E_0 - \gamma_e \langle \mathbf{L}.\mathbf{B} \rangle. \tag{5.16}$$

The Zeeman energy is determined by the expectation value of $\hat{\mathbf{L}}.\hat{\mathbf{B}}$ and when the field lies in the z-direction $\mathbf{L}.\mathbf{B} = L_z B_z$ and therefore

$$E_{\text{Zeeman}} = -\gamma_e B_z \int \Psi^* \hat{L}_z \Psi d\tau = -\gamma_e B_z \langle L_z \rangle. \tag{5.17}$$

But the expectation value of $\hat{L}_z$ is just $M_L \hbar$ and therefore

$$E_{\text{Zeeman}} = -\gamma_e B M_L \hbar. \tag{5.18}$$

The state with the highest value of M_L has the highest energy in the field, because γ_e is negative. Note that this Zeeman energy is independent of the quantum number L and even independent of the atom itself (γ_e and $\hbar$ are both constants). However, small atom-dependent and term-dependent deviations from the predicted splittings are commonly found due to the perturbation of the wavefunction by the field which has been ignored here.

Consider the ^{1}D – ^{1}P transition illustrated in Fig. 5.9. All transitions with a given value of ΔM_J ($= \Delta M_L$) will have the same energy, due to the equal level splittings in the lower and upper states. Using a mixture of parallel and perpendicular polarizations of the light ($\Delta M_J = 0, \pm 1$) the original single transition is split into three observed lines, one being a superposition of all $\Delta M_J = 0$ transitions, and one each for $\Delta M_J = \pm 1$. This triplet of lines will be observed for all types of normal Zeeman effect transitions; an example illustrated in Fig. 5.10 is the ^{1}D → ^{1}P transition of cadmium.

Fig. 5.9 Energy level diagram illustrating the normal Zeeman effect for the ^{1}D → ^{1}P transition.

(a) No field

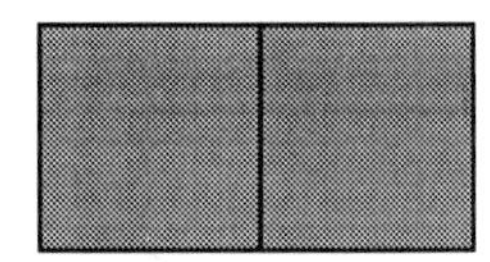

(b) With field

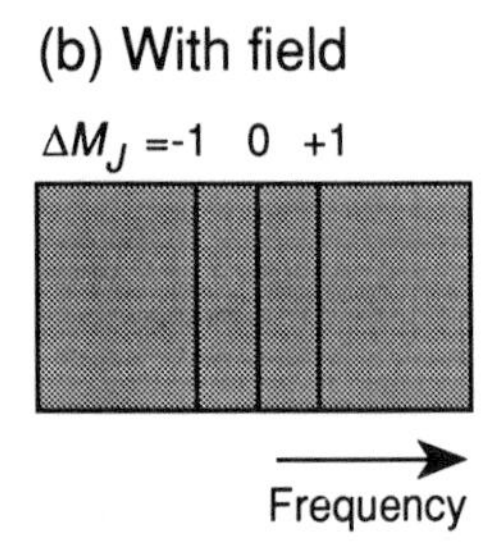

Fig. 5.10 Spectrum showing the normal Zeeman effect for the ^{1}D → ^{1}P transition of cadmium.

The anomalous Zeeman effect

For states that have both spin and orbital angular momentum, the spin–orbit coupling leads to much more complex splitting patterns in the Zeeman spectrum as illustrated in Figs 5.7a and 5.7b. Assuming the validity of Russell–Saunders coupling, the total magnetic moment **m** of the atom in a state described by quantum numbers L, S, J, M_J becomes

$$\mathbf{m} = \gamma_e(\mathbf{L} + 2\mathbf{S}), \tag{5.19}$$

(the factor of 2 comes from the g-value of the electron—see eqn 2.46), and therefore for a field along the z-direction, the term to be added to the Hamiltonian operator is given by

$$\hat{\mathcal{H}}_{\text{Zeeman}} = -\gamma_e(\hat{\mathbf{L}} + 2\hat{\mathbf{S}}).\hat{\mathbf{B}} \tag{5.20}$$

$$E_{\text{Zeeman}} = -\gamma_e B_z\langle L_z + 2S_z\rangle. \tag{5.21}$$

The expectation value of $\hat{L}_z + 2\hat{S}_z$ is not simply $(M_L + 2M_S)\hbar$ because the wavefunction, in the Russell–Saunders coupling scheme, is not an eigenfunction of the operators $\hat{L}_z$ or $\hat{S}_z$, and the quantum numbers M_S and M_L can no longer be precisely specified; we can only know that their sum $M_L + M_S$ is equal to M_J. We therefore need to derive an expression which depends explicitly on M_J rather than M_S and M_L.

Let us assume that the Zeeman part of the Hamiltonian operator can be re-expressed in the form;

$$\hat{\mathcal{H}}_{\text{Zeeman}} = -\gamma_e g(\hat{\mathbf{J}}.\hat{\mathbf{B}}) = -\hat{\mathbf{m}}.\hat{\mathbf{B}}, \tag{5.22}$$

where g is a constant to be determined. From Fig. 5.11 we see that the component of the magnetic moment along the direction of **J** is given by;

$$\mathbf{m}_J = \gamma_e \mathbf{J}\frac{(2S\cos\beta + L\cos\alpha)}{J}, \tag{5.23}$$

where $S = |\mathbf{S}|$ etc. But, as $S\cos\beta + L\cos\alpha = J$, eqn (5.23) can be rewritten as;

$$\mathbf{m}_J = \gamma_e \mathbf{J}\left(1 + \frac{S\cos\beta}{J}\right) \tag{5.24}$$

$$= \gamma_e \mathbf{J}\left(1 + \frac{\mathbf{S}^2 + \mathbf{J}^2 - \mathbf{L}^2}{2\mathbf{J}^2}\right), \tag{5.25}$$

(as $L^2 = S^2 + J^2 - 2SJ\cos\beta$ and $S^2 = \mathbf{S}.\mathbf{S} = \mathbf{S}^2$). Substituting eqn (5.25) into eqn (5.22), approximating **m** by $\mathbf{m}_J$ gives,

$$\hat{\mathcal{H}}_{\text{Zeeman}} = -\gamma_e\left(1 + \frac{\mathbf{S}^2 + \mathbf{J}^2 - \mathbf{L}^2}{2\mathbf{J}^2}\right)\mathbf{J}.\mathbf{B}. \tag{5.26}$$

Finally, taking the expectation values of the operators $\mathbf{S}^2$, $\mathbf{J}^2$ and $\mathbf{L}^2$, and assuming the magnetic field lies along the z-direction gives the result

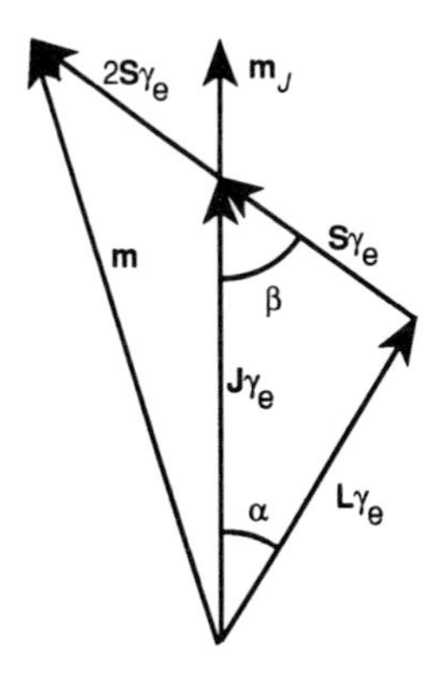

Fig. 5.11 The resultant magnetic moment **m** is equal to $\gamma_e(\mathbf{L} + 2\mathbf{S})$, while its component along the direction of $\gamma_e\mathbf{J}$ is indicated by the vector $\mathbf{m}_J$.

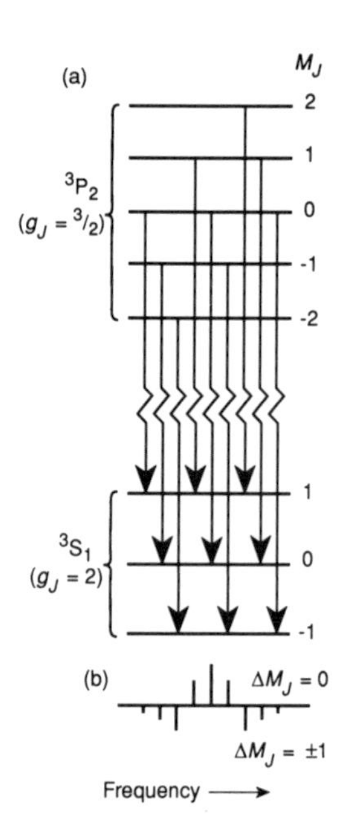

Fig. 5.12 Anomalous Zeeman effect for a $^3P_2 \rightarrow {}^3S_1$ transition, showing (a) energy levels and (b) spectrum.

$$E_{\text{Zeeman}} = -g_J \gamma_e \langle J_z B_z \rangle = -g_J \gamma_e \hbar M_J B, \tag{5.27}$$

where the *Landé g-factor* is given by

$$g_J = \frac{3J(J+1) + S(S+1) - L(L+1)}{2J(J+1)}. \tag{5.28}$$

The energy splittings now depend explicitly on the S, L and J quantum numbers for the levels concerned. Effectively, in replacing $\mathbf{m}$ by $\mathbf{m}_J$ above, we are ignoring the interaction of the field with components of the magnetic moment vector (5.19) that are perpendicular to $\mathbf{J}$ (see Fig. 5.11). Hence the expressions (5.27, 5.28) are a first-order approximation.

As an example, consider a transition of the type $^3P_2 \rightarrow {}^3S_1$.

$$\text{For } {}^3P_2, \ g_J = \frac{3 \times 6 + 2 - 2}{2 \times 6} = 1.5 \tag{5.29}$$

$$\text{For } {}^3S_1, \ g_J = \frac{3 \times 2 + 2}{2 \times 2} = 2.0. \tag{5.30}$$

The splitting of the $^3P_2$ level is smaller than for the $^3S_1$ level and therefore the set of transitions with $\Delta M_J = 0$ all have different energies as illustrated in Fig. 5.12 (cf. Fig. 5.86). In this case, nine separate lines will appear in the spectrum, including all $\Delta M_J = 0, \pm 1$ components. In examples like this the anomalous Zeeman effect is very useful because it can help to identify the term symbols of an unidentified transition. Furthermore, deviations of the observed g-factors from those predicted could be indicative that the Russell–Saunders coupling scheme is not entirely appropriate for the observed levels, or that configuration interaction is occurring.

5.6 Laser magnetic resonance

Laser magnetic resonance (LMR) is a modern spectroscopic technique which makes use of the Zeeman effect. In this case, instead of observing absorption or emission as a function of frequency for fixed magnetic field, the light frequency is fixed and the absorption is monitored as a function of the applied magnetic field strength. The aim here is to change the separation of the energy levels such that the energy differences become equal to the photon energy $h\nu$, when absorption will occur. Figure 5.13 illustrates how transitions in the $^3P_2 \leftarrow {}^3P_1$ band of the silicon atom can be successively tuned into resonance with a wavelength of 68.04 μm using a magnetic field in the range 1.083 9–1.087 2 Tesla. In this case the g_J-values for the $^3P_1$ and $^3P_2$ states, calculated from (5.28) are both 1.5. The three $\Delta M_J = +1$ transitions therefore occur at almost exactly the same field, with second-order effects, nonlinear in the field, leading to a slight separation of the lines.

The advantage of this technique is that we can make use of the very high monochromaticity and intensity provided by certain types of fixed-frequency laser, greatly enhancing the sensitivity and resolution over conventional absorption measurements. It should be noted that the transitions illustrated are

Table 5.3
Nuclear spin quantum numbers

Isotope	I	Isotope	I
^{1}H	$\frac{1}{2}$	^{15}N	$\frac{1}{2}$
^{2}H (D)	1	^{16}O	0
^{7}Li	$\frac{3}{2}$	^{17}O	$\frac{5}{2}$
^{11}B	$\frac{3}{2}$	^{19}F	$\frac{1}{2}$
^{12}C	0	^{23}Na	$\frac{3}{2}$
^{13}C	$\frac{1}{2}$	^{29}Si	$\frac{1}{2}$
^{14}N	1	^{59}Co	$\frac{7}{2}$

actually strongly forbidden, because they do not involve a change of electron configuration, hence $\Delta l = 0$. These transitions are very weakly allowed as a result of the interaction of the *magnetic* component of the electromagnetic field with the atom rather than the usual electric component. Such processes are typically several orders of magnitude weaker than normal electric dipole transitions. Their observation as illustrated in Fig. 5.13 serves to demonstrate the unusual sensitivity of the LMR method.

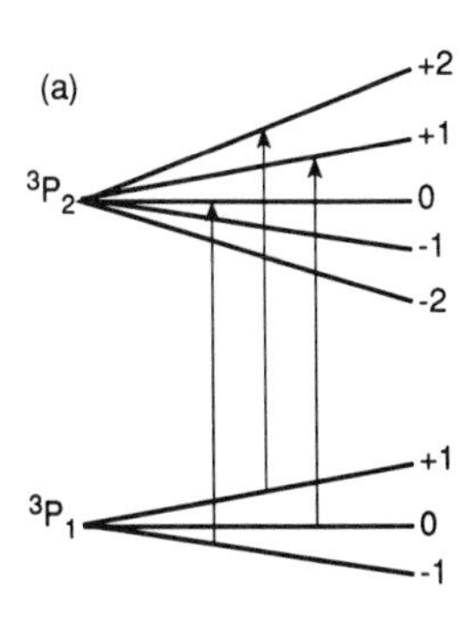

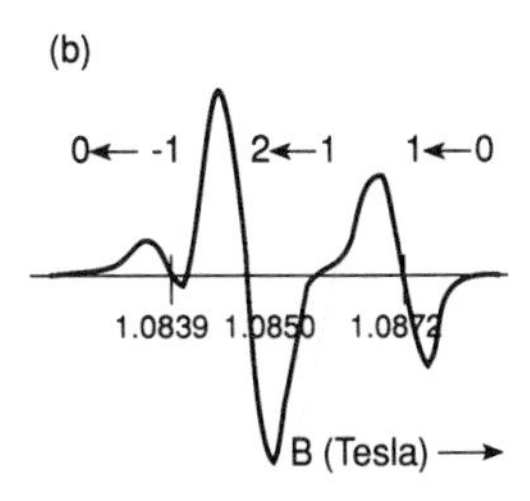

Fig. 5.13 Laser magnetic resonance transitions within the ^{3}P state of silicon. (a) Energy level diagram (schematic). (b) The three components of the $J = 2 \leftarrow 1$ transition observed using the wavelength 68.09 μm at the magnetic field strengths indicated. The spectrum shown is the first-derivative of the absorption signal, the true line-centres occurring at the null-point between each pair of positive and negative maxima.

5.7 Hyperfine structure

In Section 2.10 the idea of electrons possessing a spin angular momentum and corresponding magnetic moment was introduced, and has been shown to have important effects on the observed spectra in many cases. Some other types of elementary particle also have an intrinsic spin, including protons and neutrons. The protons and neutrons combine together to produce a nucleus with a resultant spin angular momentum, generally given the symbol **I**. Each nucleus also has an associated *nuclear spin quantum number*, I defined such that the total nuclear spin angular momentum has the quantized value

$$|\,\mathbf{I}\,| = [I(I+1)]^{\frac{1}{2}}\hbar. \tag{5.31}$$

As with other types of angular momenta, we can also define a projection quantum number M_I such that $\langle I_z \rangle = M_I\hbar$; this quantum number can take integral or half-integral values (but not both) ranging from $-I$ to $+I$. Without discussing all the details of spin coupling between protons and neutrons inside a nucleus we can state some simple rules for the possible resultant angular momentum of a nucleus as a whole.

1. For an even number of protons and even number of neutrons in a nucleus the nuclear spin quantum number is zero.
2. For an even number of neutrons and odd number of protons, or vice versa, I is half integral.
3. For odd numbers of both protons and neutrons, I is integral.

Table 5.3 lists values for I for various nucleii; note that the values are isotope-dependent.

The significance of nuclear spin in atomic spectroscopy is that the associated nuclear magnetic moment can couple to the other magnetic moments in the atom, particularly the electron spin, leading to splittings or shifts of energy levels. The magnetic dipole–dipole interaction between the nuclear and electronic spin angular momenta is nearly averaged to zero for an s electron as a consequence of the symmetrical electronic orbital motion. However, there remains a mechanism of coupling, known as the *Fermi contact interaction*, which arises from that part of the electronic wavefunction which is located *inside* the nucleus. In Section 2.3 we noted that the s electron probability density functions are nonzero at the nucleus. When the electron is in the classically forbidden region inside the nucleus, the electron–nucleus magnetic interaction

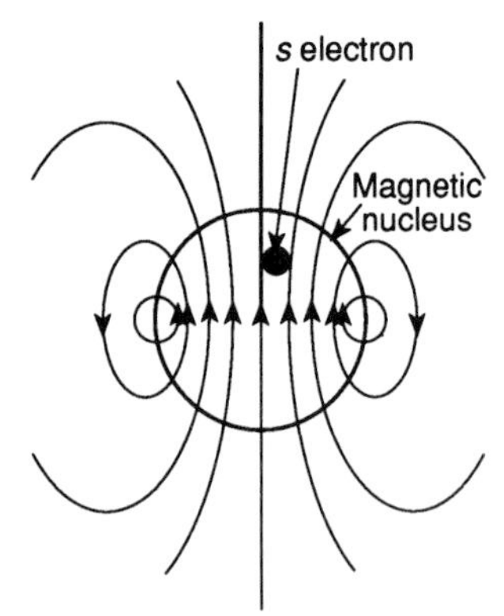

Fig. 5.14 Diagrammatic representation of the origin of the Fermi contact interaction. The field lines indicate the direction and strength of the magnetic field (the closer the lines, the stronger the field) experienced by an electron very close to or inside a magnetic nucleus. The spherical average of the field over all electron positions is not zero for the region inside the nucleus, in contrast to the region outside the immediate vicinity of the nucleus.

is not averaged to zero, as shown in Fig. 5.14. For p electrons in unfilled sub-shells the dipolar interaction outside the nucleus is not averaged to zero, but the Fermi contact interaction disappears, because there is no electron density inside the nucleus. As the Fermi contact interaction for s orbitals is generally much larger than the magnetic dipole interaction for p orbitals, the magnetic hyperfine interaction is most important for atoms that have significant unpaired electron density in an s orbital. The Fermi contact interaction is also an important mechanism for splitting of resonance lines in nuclear magnetic resonance (NMR) spectroscopy and electron spin resonance (ESR) spectroscopy.

Angular momentum coupling

With the new source of angular momentum to be considered, a new total angular momentum vector **F** is defined as

$$\mathbf{F} = \mathbf{J} + \mathbf{I}. \tag{5.32}$$

The corresponding quantum number F can take values $J + I, J + I - 1, \ldots \mid J - I \mid$. For the hydrogen atom, with the electron in an ns orbital, $J = \frac{1}{2}$ and $I = \frac{1}{2}$, hence two angular momentum states $F = 1$ or 0 are obtained, which show a small energy splitting due to the Fermi contact interaction. The $F = 1$ state is threefold degenerate ($M_F = +1, 0, -1$) and the $F = 0$ state is singly degenerate ($M_F = 0$ only). In an np state there is no Fermi contact interaction and therefore the hyperfine splitting of the ${}^2\mathrm{P}_{\frac{1}{2}}$ or ${}^2\mathrm{P}_{\frac{3}{2}}$ states is rather smaller. The energy levels for the $n = 1, 2$ and 3 states of the H atom with hyperfine splittings indicated are shown in Fig. 5.15a.

The selection rules now applicable for transitions are

$$\Delta F = 0, \pm 1 \qquad \Delta M_F = 0, \pm 1 \qquad \Delta J = 0, \pm 1, \tag{5.33}$$

but $F = 0 \rightarrow F = 0$ is forbidden, and the hyperfine components which in

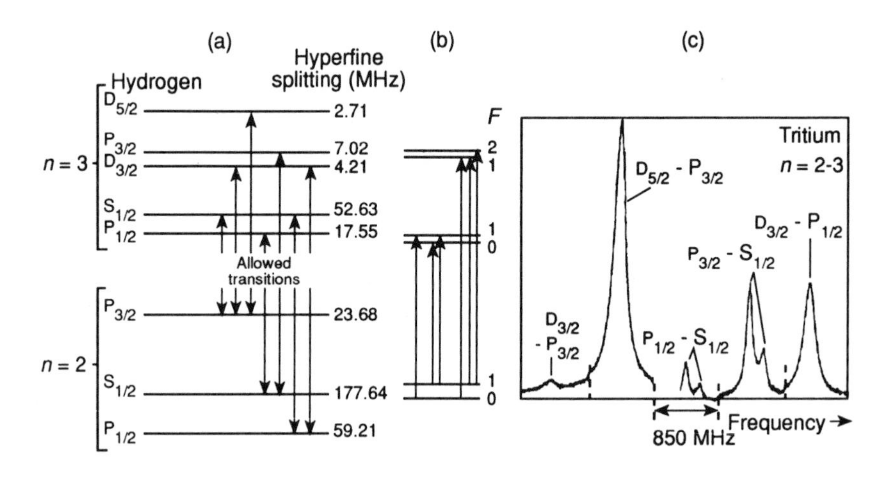

Fig. 5.15 (a) Fine and hyperfine structure in the $n = 2$ to $n = 3$ transition of hydrogen. The hyperfine splittings (MHz) are indicated at the right-hand side of each level.
(b) Permitted transitions between hyperfine components (2s $\rightarrow$ 3p only shown).
(c) Very high-resolution saturation spectrum of the tritium isotope of hydrogen, showing splittings of the components with the $2s\,{}^2\mathrm{S}_{\frac{1}{2}}$ lower state. (Three unsplit transitions related to (a) are also shown).

principle could be observed for the $3p \leftarrow 2s$ transition are illustrated in Fig. 5.15b.

In order to observe the very small hyperfine splittings in the normal emission spectrum of any element, extremely high instrumental resolution would be needed; this is particularly difficult for the Lyman series of hydrogen because the transitions lie in the vacuum ultraviolet region of the spectrum, and the splittings are a very small fraction of the actual transition frequency. In the visible region, where the Balmer series is observed, some specialist laser techniques are available for obtaining 'sub-Doppler spectra', i.e., spectra in which the normal Doppler effect contribution to the linewidth, which would obscure the hyperfine structure, has been eliminated. The reader is referred to the Background Reading for further details of the experimental methods, but Fig. 5.15c shows such a spectrum for the $3p \leftarrow 2s$ line of tritium, the mass three isotope of hydrogen, revealing the hyperfine splitting of the $2s$ level; the 3p splitting is too small to have any effect on the spectrum.

The observation of hyperfine structure may seem a little esoteric, but in fact the magnitude of the observed splittings for many-electron atoms, could be important in determining the electron configuration. A large hyperfine splitting implies unpaired electron density in an s orbital. In some cases details of the configuration mixing of the excited state might be determined.

5.8 The atomic clock

A practical example of the value of hyperfine splittings is in the definition of frequency standards via the use of *atomic clocks*. The unit of time, the second, is currently defined with respect to the caesium atomic clock standard, namely, it is the length of time taken for the radiation which is in exact resonance with a particular hyperfine transition of caesium to undergo exactly 9 192 631 770 cycles. To turn the definition on its head, the frequency of that transition in caesium is defined to be exactly equal to 9 192 631 770 cycles per second. The nuclear spin quantum number of Cs is $I = \frac{7}{2}$ and this spin couples with the unpaired electron spin $S = \frac{1}{2}$ in the $6s$ ground state to give two states $F = 4$ and 3. The atomic frequency standard involves a transition directly between these two states; the transition is forbidden by the electric dipole selection rules, because there is no change of l, but it is weakly induced by the magnetic part of the electromagnetic field.

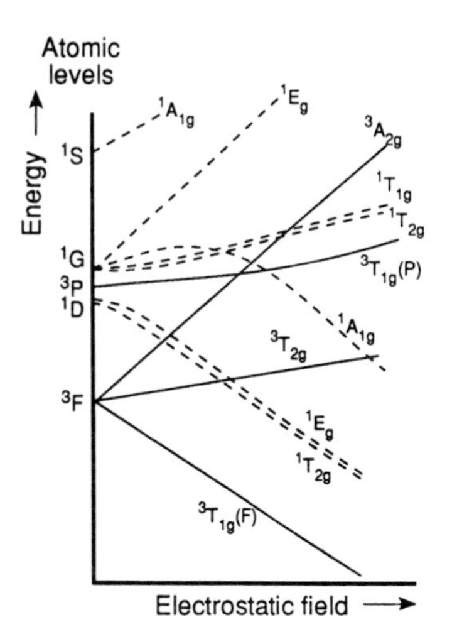

Fig. 5.16 Orgel diagram showing the perturbation of the atomic energy levels of a d^2 ion (e.g., V^{3+}) by an octahedral electrostatic field. The levels are given the appropriate symmetry label in the octahedral O_h group.

5.9 The transition metals

An important aspect of transition metal chemistry is the ability of the atoms or, more importantly, their positive ions to form complexes in solution or in the crystalline phase with a wide variety of ligands, e.g., $[Fe(H_2O)_6]^{3+}$, $[Cr(C_2O_4)_3]^{3-}$. The electronic structure of these complexes can, in many cases, be understood by use of the so-called *crystal field model*. In the weak-field limit of this model, the electronic structure of the naked transition metal ion is of dominant importance, but is perturbed by electrostatic effects due to the surrounding ligands. Ligand–metal chemical bonding is presumed to be relatively unimportant. Figure 5.16 illustrates the splitting of atomic energy levels of an ion with a d^2

configuration (see below), caused by the internal Stark effect due to an electrostatic ligand field of increasing magnitude from left to right. The atomic spectra of the transition metals and their ions in the gas phase form a valuable starting point for a more general understanding of transition metal chemistry.

Experimental spectroscopy of refractory species

Obviously the transition metals do not naturally exist in the gas phase and therefore special methods must be used to obtain spectra. One useful method is the hollow cathode discharge, discussed in Section 3.1 (see Fig. 3.3), in which the cathode is made out of the transition metal of interest. Transition metal atoms are brought into the gas phase by the impact of highly energetic positive ions at the cathode. The hollow cathode is likely to be a good source for emission spectra. An alternative is the microwave-excited electrodeless discharge, in which a sealed glass tube containing a carrier gas (e.g., argon) and a transition metal halide is exposed to intense microwave radiation. The oscillating electric field of the radiation accelerates any free electrons or ions leading to a breakdown of the carrier gas, and then heating effects in the discharge lead to vaporization of the halide. The molecules are then dissociated and the resultant atoms are excited in the discharge, leading to emission.

For measurements of *absorption* spectra it is desirable to achieve vaporization without excitation, and a common method is to use the King's Furnace, in which a tantalum tube containing the sample is inductively heated by a surrounding coil, providing a uniform temperature distribution over a fairly large volume (in contrast to the temperature distribution of a flame).

Electron configurations

The transition metal atoms are characterized by having a number of unpaired d electrons in their ground and excited electronic configurations. In the atom preceding the first transition series, calcium, the outermost electrons occupy the $4s$ orbital in preference to the $3d$ orbitals (ground state configuration $1s^2\,2s^2\,2p^6\,3s^2\,3p^6\,4s^2$). This energetic preference for the orbital of higher principal quantum number is a consequence of the penetration and shielding effects, referred to in Section 3.5; the more penetrating $4s$ orbital experiences a higher effective nuclear charge than the $3d$ electron, and despite its higher principal quantum number has lower energy. For the first few elements of the first transition series (ground configurations listed in Table 5.4) the ground state configuration $\ldots 4s^2\,3d^{n-2}$ is adopted (n is the number of electrons outside the $\ldots 3s^2\,3p^6$ core). However, the lowest terms of the configurations $4s^1\,3d^{n-1}$ and $4s^0\,3d^n$ are also quite close in energy as illustrated in Fig. 5.17. The energy gap decreases until at Cr the $4s\,3d^5$ configuration is lower in energy than the $4s^2\,3d^4$ configuration. We are observing here the well-known contraction of the $3d$ orbitals as the atomic number increases. The explanation is that as each $3d$ electron is added, it does very little to shield the other $3d$ electrons from the increased nuclear charge; each d electron can be added to a different orbital with a different spatial distribution and with parallel spin, minimizing electron–electron repulsions. Thus, the effective nuclear charge experienced by the d orbitals increases rather dramatically compared to the increase for the $4s$ orbital,

Table 5.4

Term symbols and configurations for first-row transition metal atoms

Element	Configuration	Terms arising	Ground state level
Sc	$[Ar]\,4s^2\,3d$	^2D	$^2D_{3/2}$
Ti	$[Ar]\,4s^2\,3d^2$	$^1(SDG)\,^3(PF)$	$^3F_2$
V	$[Ar]\,4s^2\,3d^3$	$^2(D)\,^2(PDFGH)\,^4(PF)$	$^4F_{3/2}$
Cr	$[Ar]\,4s^1\,3d^5$	$^{1,3}(D)\,^{1,3}(PDFGH)\,^{3,5}(PF)\,^{1,3}(SDFGI)\,^{3,5}(DG)\,^{5,7}(S)$	$^7S_3$
Mn	$[Ar]\,4s^2\,3d^5$	$^2(D)\,^2(PDFGH)\,^4(PF)\,^2(SDFGI)\,^4(DG)\,^6(S)$	$^6S_{5/2}$
Fe	$[Ar]\,4s^2\,3d^6$	$^1(SDG)\,^3(PF)\,^1(SDFGI)\,^3(PDFGH)\,^5(D)$	$^5D_4$
Co	$[Ar]\,4s^2\,3d^7$	$^2(D)\,^2(PDFGH)\,^4(PF)$	$^4F_{9/2}$
Ni	$[Ar]\,4s^2\,3d^8$	$^1(SDG)\,^3(PF)$	$^3F_4$
Cu	$[Ar]\,4s^1\,3d^{10}$	^2S	$^2S_{1/2}$
Zn	$[Ar]\,4s^2\,3d^{10}$	^1S	$^1S_0$

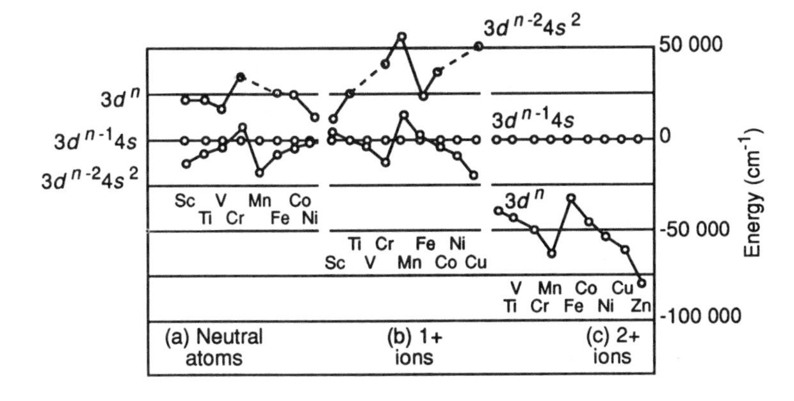

Fig. 5.17 The energies of the lowest energy terms of the low-lying electron configurations of the first transition series, relative to the $3d^{n-1}4s$ configuration. (a) Neutral atoms. (b) Monovalent (1+) ions. (c) Divalent (2+) ions.

and the $3d$ orbital energy decreases more rapidly. It should be emphasized that there is no 'special stability' associated with a half-full shell of d electrons; the apparent stability of the $4s^13d^5$ configuration is simply the end-result of a progressive trend. The trend is abruptly halted at the next element because the configuration $4s^13d^6$ would have two electrons in the same $3d$ orbital; in that case, the shielding would be much more effective as the two electrons must be relatively close spatially. The two electrons in the spherical $4s$ orbital experience less mutual repulsion, hence the configuration $4s^23d^5$ is lower in energy and is the ground state for Mn. The $3d$ contraction continues across the periodic table and eventually this effect wins over again at copper which adopts the $3d^{10}4s^1$ configuration.

The relative energies of the lowest terms of various configurations of the monovalent and divalent ions are also shown in Fig. 5.17. The $3d^{n-2}4s^2$ configuration is now the highest in energy of the three, with either the $3d^{n-1}4s$ or $3d^n$ configurations being most stable. For the 2^+ ions, the $3d^n$ configuration is always most stable. The higher effective nuclear charge for the positive ions relative to the neutral atoms, leads to a significant contraction of the d orbitals, hence the enhanced preference for $3d$ orbital occupation.

The spectra of the atomic transition metals

The general situation is that there will be a high density of low-lying electronic states arising from all three configurations $4s^2\,3d^{n-2}$, $4s^1\,3d^{n-1}$ and $4s^0 3d^n$ especially when all the terms and levels arising from these configurations are considered. In view of the methods used to produce the atoms or ions in the gas phase, it is probable that a wide range of these low-lying states will be populated and therefore the absorption spectra will be very complicated. Furthermore the quantum mechanical interpretation of the spectra may be confused by a substantial degree of *configuration interaction* (see Section 4.7). Transitions *between* these lowest configurations are forbidden because they would involve $d \rightarrow s$ or vice versa for which $\Delta l = \pm 2$. Therefore, the lowest fully-allowed transitions observed will involve excitation of either a $3d$ or a $4s$ electron to a $4p$ orbital.

The terms arising from the various ground state configurations of the first-row transition metals are given in Table 5.4; these can be derived by extension of the Russell–Saunders vector coupling arguments given in Section 5.2, with careful consideration of the Pauli principle. Russell–Saunders coupling is generally found to be a good approximation for the atoms of the first transition series; applying Hund's rules, the ground states are as listed. The transitions should obey the normal selection rules

$$\Delta J = 0 \pm 1, \;\; \Delta L = 0, \pm 1, \;\; \Delta l = \pm 1, \;\; \Delta S = 0. \qquad (5.34)$$

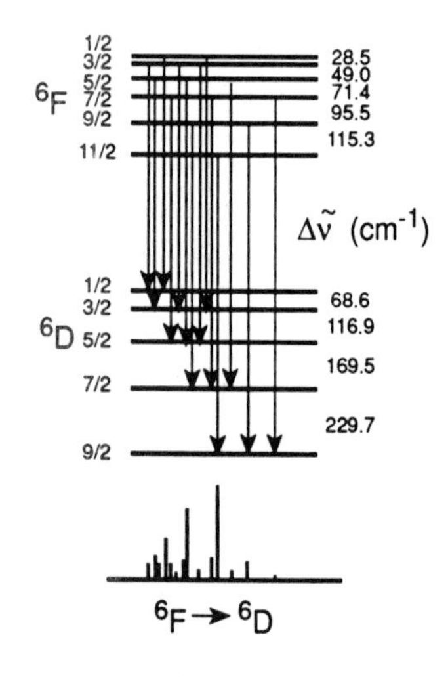

Fig. 5.18 The ^{6}F $\rightarrow$ ^{6}D transition of manganese is shown to split into 14 components through spin–orbit coupling.

Figure 5.18 illustrates some of the transitions that have been observed in the manganese atom in the wavelength region around 380 nm. The transition involves the $3d^6 4s$ (^{6}D) and $3d^6 4p$ (^{6}F) terms, and the spin–orbit components (five for ^{6}D and six for ^{6}F) are illustrated. The Landé interval rule, eqn (4.14), is approximately obeyed.

Crystal field effects

Many transition metal *complexes* in solution are coloured, indicating transitions to relatively low-lying electronically-excited states. The transitions are fairly weak, and can generally be ascribed, on an atomic model, to transitions between different states arising from the same configuration. These *d–d* transitions are formally forbidden in the atom, but the symmetry breaking perturbation due to the ligand field leads to an admixture of configurations with $4p$ character, into the $4s^m 3d^{n-m}$ configurations, and a partially allowed transition becomes possible. Even when the ion is in a centrosymmetric environment, so that the inversion symmetry is retained, a *d–d* transition can become allowed if accompanied by the excitation of a nonsymmetric vibration. Nevertheless, some of the atomic selection rules still apply, particularly the $\Delta S = 0$ rule. Thus, Mn^{2+} or Fe^{3+} complexes are almost colourless, because the ground state of these ions, configuration $3d^5 4s^0$, has the term ^6S. This is the only sextet state ($S = \frac{5}{2}$) arising from the ground configuration and consequently there are no spin-allowed transitions within this configuration. For the d^2 and d^8 ions, (see Fig. 5.16), the transitions observed in the spectra of various complexes (e.g., $[V(H_2O)_6]^{3+}$ or $[Ni(H_2O)_6]^{2+}$ are predominantly of $^3P \leftarrow {}^3F$ and $^3F \leftarrow {}^3F$ character in accord with the $\Delta S = 0$ selection rule of the atomic spectrum, but not fully in keeping with the $\Delta L = 0, \pm 1$ selection rule.

5.10 Photoelectron spectroscopy of atoms

Photoelectron spectroscopy (PES) is somewhat different from conventional absorption or emission spectroscopy in that the excitation process observed is between the atomic (or molecular) ground state and the ionization continuum. The method was invented in the early 1960s in independent experiments by Turner, Siegbahn and Vilesov, and a typical apparatus used is illustrated in Fig. 5.19a. Instead of monitoring the absorption of light as a function of frequency, a monochromatic fixed-frequency lamp is used to ionize the atoms, and the spectrum of kinetic energies of the ejected photoelectrons is measured. In Fig. 5.19a the voltage applied to the curved electrostatic sector is such that only electrons of one particular energy are transmitted to the detector; this voltage can be varied, changing the required kinetic energy for transmission.

Above the ionization limit, a continuum of energy states exist for the outermost electron in an atom; the electron is free from the nucleus and its energy levels are not quantized. Absorption leading to ionization will occur for any energy greater than the ionization energy. As a result of conservation of energy, the input photon energy, $h\nu$ must equal the adiabatic ionization energy (I) plus the kinetic energy of the ion and electron, plus any internal energy of the ion, i.e., electronic energy over and above its normal ground state energy (see Fig. 5.19b)

$$h\nu = I + E_{\mathrm{el}} + E_{\mathrm{ion}}. \tag{5.35}$$

Conservation of momentum, and the very light mass of the electron compared to the ion means that nearly all the kinetic energy belongs to the electron, hence E_{ion} is just equal to the ionic *internal* energy.

In the most simple case, the hydrogen atom, the electrons must all have the same kinetic energy; the ion cannot have internal energy because it has no electrons, hence $E_{\mathrm{el}} = h\nu - I$. In a typical experiment a helium lamp is used providing photons of energy $h\nu = 21.2\,\mathrm{eV}$, ($1\,\mathrm{eV} \equiv 8065.5\ \mathrm{cm}^{-1}$) and for hydrogen the kinetic energy spectrum shows a single peak at $E_{\mathrm{el}} = 7.6\,\mathrm{eV}$ indicating that the ionization potential is 13.6 eV. Thus, we see that photoelectron spectroscopy is a useful method for determining ionization potentials. However, for more complicated atoms much more information is obtained.

Figure 5.20 shows the spectrum of electron kinetic energies for the rare gas atoms, using an excitation energy $h\nu = 40.81$ eV; in each case the two peaks must be ascribed to the formation of two different internal energy states of the ion. The ground state of argon has the configuration $1s^2\,2s^2\ \ldots 3p^6$, and removal of the outermost electron gives an ion with configuration $\ldots 3p^5$. The ion is isoelectronic with the chlorine atom and by analogy we would expect

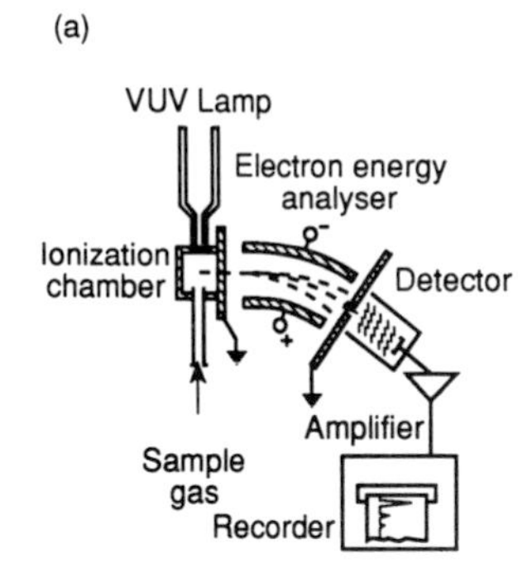

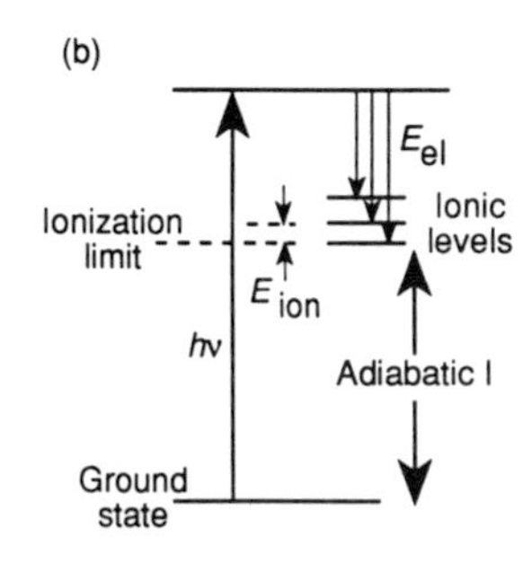

Fig. 5.19 (a) Schematic diagram of a photoelectron spectrometer. The lamp provides monochromatic vacuum ultraviolet radiation to ionize the gas sample, and the electrons are dispersed by the curved electrostatic sector according to their kinetic energy.
(b) Energetics of photoelectron spectroscopy, showing the ground state energy level of the neutral molecule and a set of energy levels for the ion; the input photon energy $h\nu$ less the adiabatic ionization energy must equal the combined energies of the photoelectron and photoion.

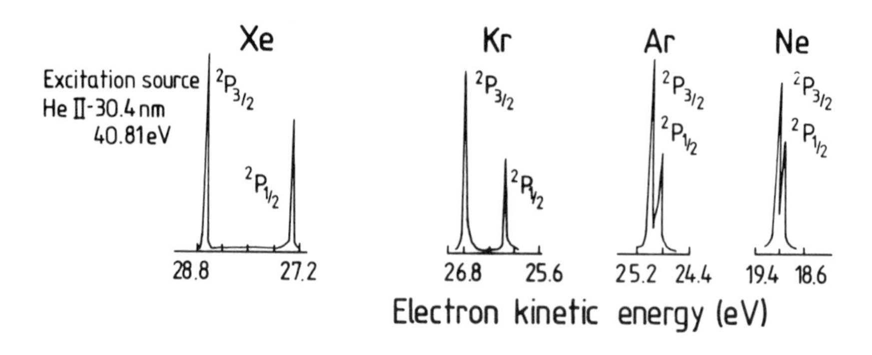

Fig. 5.20 The photoelectron spectra of the rare-gas atoms recorded using the He(II) photon source.

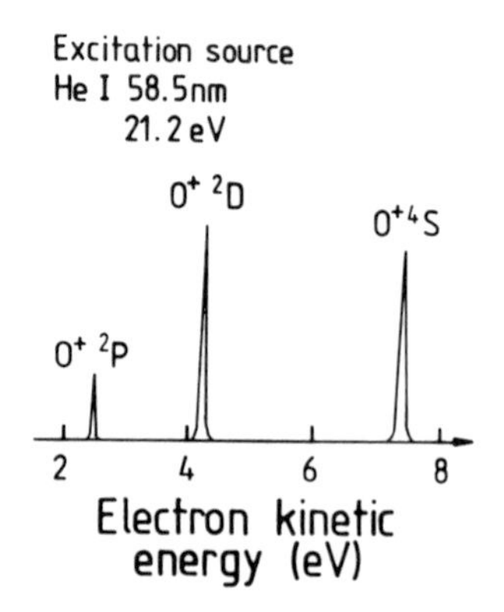

Fig. 5.21 The photoelectron spectrum of the oxygen atom using the He(I) photon source. The atoms were produced in a microwave discharge and the peaks that were also observed due to molecular oxygen have been subtracted from the signal.

there to be two spin orbit states $^2P_{\frac{1}{2}}$ and $^2P_{\frac{3}{2}}$ associated with this configuration. If the ionization leads to the lower state $J = \frac{3}{2}$, this corresponds to the *adiabatic* ionization limit for which the ion has no internal energy. Consequently $E_{el} = h\nu - I$. However, if the upper $J = \frac{1}{2}$ state is produced then $E_{el} = h\nu - I - E_{ion}$ where E_{ion} is equal to the spin–orbit splitting of the ionic ^2P state. Thus, it can be seen that the photoelectron spectrum maps out the quantum states of the ion that can be produced by removal of a single electron from the corresponding neutral atom. This becomes more clear looking at the photoelectron spectrum of atomic oxygen. The oxygen ion O^+ has the ground state configuration $1s^22s^22p^3$ and is isoelectronic with the nitrogen atom. The possible ionic states arising from this configuration are ^4S, ^2D and ^2P and these three peaks are seen in the photoelectron spectrum as illustrated in Fig. 5.21. The photoelectron kinetic energy differences in the spectrum are equal to the energy differences between these states. In this case the spin–orbit splittings are too small to be observed.

The relative intensities of the two peaks in the rare gas spectra are 2:1, reflecting the ratios of the degeneracies of the ionic states. The $J = \frac{3}{2}$ state is fourfold degenerate with M_J components $\pm\frac{3}{2}, \pm\frac{1}{2}$, whereas the $J = \frac{1}{2}$ state is doubly-degenerate. Assuming the probability of forming any ionic M_J state by photoionization is equal, the ratio for $J = \frac{3}{2}$ to $J = \frac{1}{2}$ should be 4:2, as is observed.

If more energetic radiation is used for ionization, e.g., the magnesium X-ray line at 1254 eV, then more peaks can be observed in the photoelectron spectra, as illustrated for argon in Fig. 5.22. The electrons of lowest

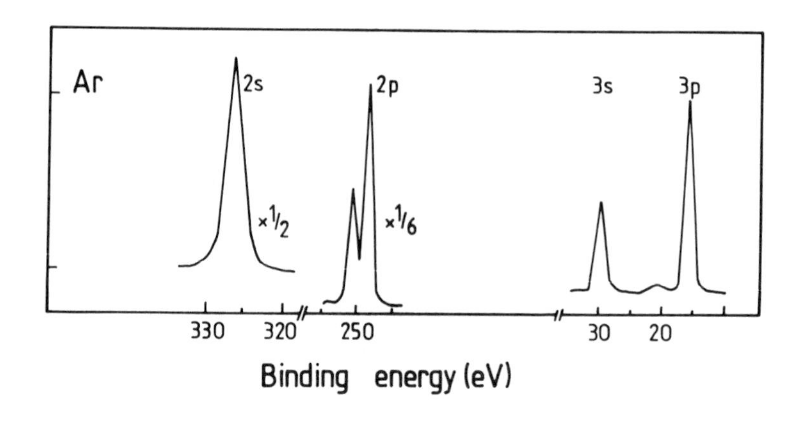

Fig. 5.22 X-ray photoelectron spectrum of argon recorded using the magnesium line at 1254 eV. The horizontal scale, binding energy, gives the energy required to remove the electron to form a specified state of the ion, and was obtained by subtracting the observed photoelectron energy from the photon energy.

kinetic energy are ejected from the more tightly bound orbitals; removal of a $2s$ electron in argon leaves the configuration $2s\,2p^6\,3s^2\,3p^6$ giving the lowest electron-energy peak (highest binding energy) in the figure. Some weak 'satellite' peaks can also be observed between the intense main peaks of the spectrum. These can be ascribed to two-electron processes involving the ionization of one electron and the simultaneous excitation of a second, e.g., the production of a state $3s^2\,3p^4\,3d$ in the argon ion, in which one $3p$ electron is removed and one $3p$ electron is excited to the $3d$ orbital. Such processes have their origins in electron correlation as discussed in Section 4.7.

Koopmans' theorem

The difference between the $3s$ and $3p$ electron energies shown in Fig. 5.22 is 13.5 eV which must equal the difference of energy between the ionic configurations $3s\,3p^6$ and $3s^2\,3p^5$, ignoring spin–orbit coupling (which is not resolved at the resolution of the spectrum shown). It is therefore equal to the $3s \rightarrow 3p$ transition energy in the ion, that could also be observed by direct emission or absorption spectroscopy on the ion. It is also equal to the difference of the ionization energies of a $3s$ or $3p$ electron in the neutral atom.

Koopmans' theorem states that this ionization energy difference is approximately equal to the difference of the *orbital energies* for the *neutral atom*. The $3p$ orbital energy is the one-electron energy that we would calculate by solving the Schrödinger equation for one of the $3p$ electrons in the electrostatic potential due to the averaged distribution of all other electrons present. This energy would normally be calculated using the self-consistent field (SCF) approach, discussed in Section 3.2. In order for Koopmans' theorem to be valid, it would be required that the wavefunctions of the other electrons do not change when the ionized electron is removed. In reality, there will always be some electronic reorganization following ionization which results in an energetic payback; consequently the ionization energy is not simply equal to the negative of the orbital energy. Nevertheless the viewpoint of Koopmans' theorem is a useful one, in that it can be thought that the photoelectron energies, as in Fig. 5.22, for example, give a crude mapping of the relative energies of the orbitals in the neutral atom.

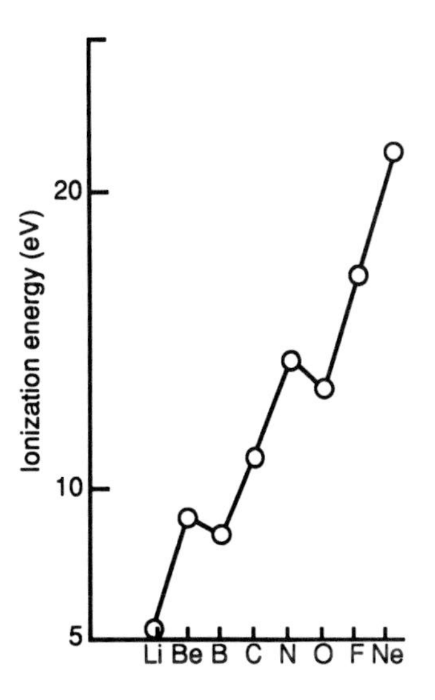

Fig. 5.23 Adiabatic ionization energies of the first-row atoms.

Ionization energies of the first-row atoms

Figure 5.23 shows the measured adiabatic ionization energies of the first-row atoms and it is worth considering the observed trends in detail. The first point to note is that the ionization energy depends not only on the electronic structure of the neutral atom but also on the electronic structure of the ion. The ionization energy is very clearly defined as an *energy difference* between the ground state of the atom and the ground state of the ion. In either case, the energy within the orbital approximation, obtained by an extension of the arguments in Section 4.6, is equal to

$$E = \sum_k h_k + \sum_{k>l} J_{kl} - \sum_{k>l} K_{kl}, \tag{5.36}$$

where $\sum h_k$ is the sum of the energies of each electron in the presence of the nucleus alone, J_{kl} and K_{kl} are the exchange and coulomb integrals (see Section 4.6) between all pairs of electrons, but K_{kl} vanishes unless the two electrons have parallel spins. Any effect which stabilizes the ionic energy will lower the ionization energy.

The overall trend is for an increase of ionization energy across the period. The explanation for this is simply that each successive electron which is added ineffectively shields the other electrons from the increasing nuclear charge, and therefore the electronic binding energy increases. The step down between beryllium and boron is due to the change from removing a $2s$ electron to removing a less-penetrating $2p$ electron. The other deviations from this overall upward trend are mainly due to spin pairing and correlation effects; thus the nitrogen atom in its ground state has the term ^{4}S with three unpaired electrons in different orbitals. Addition of a further $2p$ electron as in oxygen can only result in spin pairing with two electrons in the same orbital meaning greater electrostatic repulsion. This extra electron is therefore more easily removed to alleviate the extra repulsion and the ionization energy for oxygen is lower than nitrogen.

A simplistic calculation to illustrate the exchange energy contribution to the ionization energies of the first-row atoms uses the assumption that the exchange energy is equal to the number of pairs of spin parallel $n = 2$ electrons, times a constant $-K$ (i.e., it is assumed all the K_{kl} are equal in eqn 5.36). For the $2s^2\,2p^3\ ^4$S ground state configuration of O^+, there are four electrons with α spin and one with β spin, and consequently there are six pairs of spin-parallel electrons (from permutations of pairs of the α-spin electrons). For the $2s^2\,2p^4(^3\mathrm{P})$ ground state of the oxygen atom, there are four electrons with α spin, and two with β spin, and hence there are seven pairs of spin-parallel electrons. The decrease in exchange energy on ionization of an oxygen atom is therefore $7K - 6K = K$. Similar calculations for the whole series Li to Ne give values $0, 0, K, 2K, 3K, K, 2K, 3K$ respectively. The drop in ionization potential between nitrogen and oxygen is partly explained by the much smaller loss of exchange energy upon ionization of the latter (K) compared to the former ($3K$).

5.11 Rydberg states and autoionization

The Rydberg quantum defect formula (3.10), which can be used for 'one-electron atoms' such as the alkali metals, is not really applicable to the low-lying electronic states of many-electron atoms, because the effects of electron–electron repulsion and spin correlation lead to very large splittings of configuration energies in the Russell–Saunders coupling scheme. If, on the other hand, we excite one electron into a state of high principal quantum number, e.g., $n = 10$, the outermost electron becomes so distant from the other 'core' electrons that effects such as spin correlation and orbital angular momentum coupling become rather small, and we obtain once again a pseudo one-electron atom; in this case we describe the atom as being in a *Rydberg* state and the electron is in a *Rydberg orbital*.

The major difference between this situation and that of the alkali metals is that the atom may have an open-shell inner core of electrons; for example, the Rydberg state ... $3s^2\,3p^5\,10s$ in argon has a vacancy in the $3p$ subshell. In this case, and many others, the jj-coupling scheme is probably most appropriate. There is a core total angular momentum $\mathbf{j}_1$, with the corresponding quantum number j_1 which for the configuration ... $3p^5$ can take values $\frac{1}{2}$ or $\frac{3}{2}$, and a Rydberg electron angular momentum, which for a $10s$ orbital is $j_2 = \frac{1}{2}$ only. $\mathbf{j}_1$ and $\mathbf{j}_2$ can couple together to give total angular momentum states $J = 2, 1$ or 0, designated $[\frac{1}{2}\frac{1}{2}]_1$, $[\frac{1}{2}\frac{1}{2}]_0$, $[\frac{3}{2}\frac{1}{2}]_1$, and $[\frac{3}{2}\frac{1}{2}]_2$, using the notation $[j_1\,j_2]_J$. There will be a clear energy splitting between the states with $j_1 = \frac{1}{2}$ and $j_1 = \frac{3}{2}$ as illustrated schematically in Fig. 5.24. In Section 5.10 we saw that the photoelectron spectra of the rare gases show two peaks due to the ionization to form the ion in two possible spin–orbit states. Defining the two ionization energies, $I_{\frac{1}{2}}$ and $I_{\frac{3}{2}}$ as the energy required to produce the ions in these two states at threshold, it is found that the Rydberg energy levels of the neutral rare gas atoms, relative to the ground state energy, can all be fitted to one of two Rydberg formulae

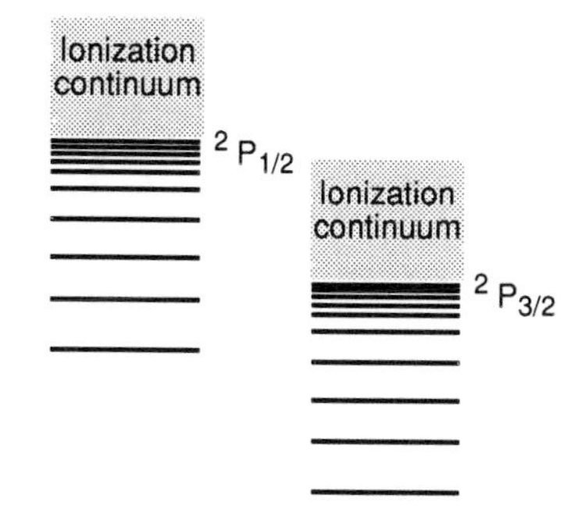

Fig. 5.24 Schematic diagram showing separate Rydberg series converging to the $^2P_{\frac{1}{2}}$ and $^2P_{\frac{3}{2}}$ ionization thresholds of a rare-gas atom.

$$E_{nlJ} = I_{\frac{1}{2}} - \frac{R}{(n-\delta_{lJ})^2} \tag{5.37}$$

$$E'_{nlJ} = I_{\frac{3}{2}} - \frac{R}{(n-\delta'_{lJ})^2}, \tag{5.38}$$

depending on whether j_1 is $\frac{1}{2}$ or $\frac{3}{2}$ respectively. There are therefore two types of interleaved Rydberg series. As with the alkali metals, Rydberg orbitals of different l have different quantum defects as a reflection of the different penetrating properties. The subscript J in the quantum defect δ_{lJ} indicates that the value also depends on the total angular momentum in the jj-coupling scheme.

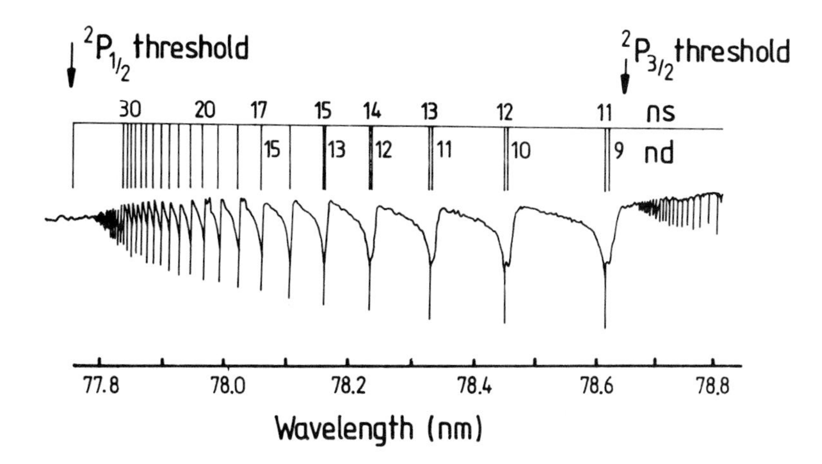

Fig. 5.25 The absorption spectrum of the argon Rydberg states. The $^2P_{\frac{1}{2}}$ and $^2P_{\frac{3}{2}}$ ionization thresholds are indicated and the *ns* and *nd* series converging to the upper ($j = \frac{1}{2}$) limit are labelled. The broad asymmetric lines are the transitions to *nd* states whereas the sharp spikes on the top of each broad line are the *ns* transitions.

Figure 5.25 shows the spectrum of the argon Rydberg states, in which the convergence of the upper states to two separate limits, $j_1 = \frac{1}{2}$ and $j_1 = \frac{3}{2}$ is clearly observed. The selection rules allow excitation of the $3p$ electron in the ground state configuration to either an ns or nd orbital ($\Delta l = \pm 1$).

The selection rule, $\Delta J = 0, \pm 1$ (but $J = 0 \rightarrow J = 0$ forbidden) restricts excitation from the $J = 0$ ground state to states of total angular momentum quantum number $J = 1$. In jj-coupling, the ns and nd states with $J = 1$ are $ns[\frac{1}{2}\frac{1}{2}]_1$ $ns[\frac{3}{2}\frac{1}{2}]_1$ $nd[\frac{1}{2}\frac{3}{2}]_1$ $nd[\frac{3}{2}\frac{5}{2}]_1$ and $nd[\frac{3}{2}\frac{3}{2}]_1$, and five Rydberg series are observed, three converging to the lower ionization limit and two to the upper limit; the two series converging to the upper limit are labelled in the figure.

An interesting phenomenon occurring between the two limits is the broadening of the lines in the spectra, particularly those involving excitation of the nd states; this is due to a process known as *autoionization*. It can be seen from Fig. 5.24 that there are bound states converging on the $j = \frac{1}{2}$ limit that are degenerate with the continuum above the $j = \frac{3}{2}$ limit. The electron is not truly bound in these states—the states would be better described as 'pseudo-bound' or 'metastable', because there is a finite probability of an energy transfer between the inner excited core and the Rydberg electron. The transfer results in the de-excitation of the core back to the $j = \frac{3}{2}$ state and gives the electron sufficient energy to ionize. This autoionization process has a finite lifetime associated with it; in a classical picture the electron may undergo a certain number of orbits and then depart, unlike the directly ionized electron which leaves immediately. The broadening of the lines can be accounted for in terms of a reformulated version of the uncertainty principle

$$\Delta t \Delta E \sim \hbar. \tag{5.39}$$

In the current context, Δt is to be interpreted as the lifetime of the excited state and ΔE as the uncertainty in its energy which is proportional to the linewidth: the shorter the lifetime, the broader the line. The typical linewidths of $\sim 5\ \text{cm}^{-1}$ correspond to a lifetime of ~ 1 ps (1×10^{-12} s). Thus, it is illustrated that dynamical information about excited states can also be obtained from atomic spectra, as well as energetic information.

5.12 Problems

1. (a) Write down a table showing the microstates permitted by the Pauli principle for an nd^2 configuration; hence derive the permitted term symbols in Russell–Saunders coupling.

   (b) Use angular momentum coupling arguments to show that the levels derived in jj-coupling for an $np\,nf$ configuration are $[\frac{1}{2}\frac{5}{2}]_{3,2}$, $[\frac{1}{2}\frac{7}{2}]_{4,3}$, $[\frac{3}{2}\frac{7}{2}]_{5,4,3,2}$ and $[\frac{3}{2}\frac{5}{2}]_{4,3,2,1}$.

2. Analysis of the atomic spectrum of a certain atom shows that the ground term is split into three levels with relative energies 0, 77 and 231 cm^{-1}. Assuming that the splitting is due to the spin–orbit coupling interaction, represented by the operator $\frac{A}{\hbar^2}\mathbf{L}.\mathbf{S}$, derive an expression for the difference in energy between the levels with total angular momentum J and $J+1$ in terms of the quantum numbers L, S and J. Hence assign J values to the three levels and obtain a value for A. What are the possible values of L and S for this term?

3. The lowest energy configuration of the N^+ ion is $2s^2\,2p^2$ giving rise to the terms ^{3}P, ^{1}D, and ^{1}S. The first excited configuration is obtained by promoting an electron from a $2s$ to a $2p$ orbital. Use angular momentum coupling arguments to derive the states which arise from this configuration, noting that a p^3 configuration as in the neutral N atom, gives rise to terms ^{4}S, ^{2}P and ^{2}D. An allowed transition has been detected in N^+ between one of the levels of its *ground* term and one of the levels associated with the first excited configuration. The g-factors of both levels were measured to be 1.50. Assuming the validity of the Landé g-factor formula, suggest an assignment for the transition.

4. Far infrared transitions have been recorded in atomic oxygen within its ground state by the technique of laser magnetic resonance. Spectra were recorded using the laser wavenumbers are listed below in perpendicular polarization ($\Delta M_J = \pm 1$). ($\partial\tilde{\nu}/\partial B < 0$ implies that the actual transition wavenumber decreases with B).

Assignment	Laser wavenumber $\tilde{\nu}$ (cm^{-1})	Magnetic field B (mT)	
$^3P_1(M_J = 1) \rightarrow {}^3P_0$	68.652 221	91.7	$\partial\tilde{\nu}/\partial B < 0$
$^3P_2(M_J = 0) \rightarrow {}^3P_1(M_J = 1)$	158.487 673	312.6	$\partial\tilde{\nu}/\partial B > 0$

In each case the spectrum appears as a single line.

(a) Illustrate these observations with the help of an energy level diagram.

(b) Deduce the two fine-structure separations for ground state O atoms from these data, given that $g = 1.500\,93$ for $J = 1$ and $\mu_B (= \gamma_e \hbar) = 0.466\,864$ cm^{-1}T^{-1}. Are these results in accord with the expectations of simple theory?

5. Predict which lines would be observed including fine and hyperfine structure for the $2p3s$ (^{3}P) $\rightarrow$ $2p^2$ (^{3}P) transition of the ^{13}C atom (nuclear spin $I = \frac{1}{2}$). Comment on the likely relative magnitudes of the hyperfine splitting in the lower and upper states.

Background reading

Andrews D.L. (1990). *Lasers in chemistry*, 2nd edn, Springer-Verlag, Berlin.

Atkins, P.W. (1983). *Molecular quantum mechanics*, 2nd edn, Oxford University Press.

Atkins, P.W. (1991). *Quanta: a handbook of concepts*, 2nd edn, Oxford University Press.

Baggott, J. (1991). *The meaning of quantum theory*, Oxford University Press.

Benson, H. (1991). *University physics*, Wiley, New York.

Bransden B.H. and Joacham C.J. (1983). *Physics of atoms and molecules*, Longman, London.

Corney, A. (1977). *Atomic and laser spectroscopy*, Clarendon Press, Oxford.

Eland, J.H.D. (1984). *Photoelectron spectroscopy: an introduction to ultraviolet photoelectron spectroscopy in the gas phase*, 2nd edn, Butterworth, London.

Hecht, E. (1987). *Optics*, 2nd edn, Addison-Wesley, Reading, Massachusetts.

Klinkenberg, P.F.A. (1976) in *Methods of Experimental Physics*, Vol. 13, (ed. Williams, D.), Chapter 3.4, Academic Press, London.

Koplitz, B., Xu, Z., Baugh, D., Buelow, S., Häusler., D., Rice, J., *et al.* (1986). Dynamics of molecular photofragmentation. *Faraday Discuss. Chem. Soc.*, **82**, 125.

Kuhn, H.G. (1971). *Atomic spectra*, 2nd edn, Longman, London.

Wineland, D.J. and Itano, W.M. (1987). *Physics Today*. **40** (6), 34.

Index